Functional Safety for Embedded Systems

This book uses automotive embedded systems as an example to introduce functional safety assurance and safety-aware cost optimization. The book explores functional safety assurance from the perspectives of verification, enhancement, and validation.

The functional safety assurance methods implement a safe and efficient assurance system that integrates safety verification, enhancement, and validation. The assurance methods offered in this book could provide a reasonable and scientific theoretical basis for the subsequent formulation of automotive functional safety standards. The safety-aware cost optimization methods divide cost types according to the essential differences of various costs in system design and establish reasonable models based on different costs. The cost optimization methods provided in this book could give appropriate cost optimization solutions for the cost-sensitive automotive industry, thereby achieving effective cost management and control. Functional safety assurance methods and safety-aware cost optimization support each other and jointly build the architecture of functional safety design methodologies for automotive embedded systems.

The work aspires to provide a relevant reference for students, researchers, engineers, and professionals working in this area or those interested in hardware cost optimization and development cost optimization design methods based on ensuring functional safety in general.

Guoqi Xie is currently a Professor with Hunan University. He received the Ph.D. degree in computer science and engineering from Hunan University, China, in 2014. He was a postdoctoral research fellow with Nagoya University, Japan. His current research interests include real-time systems, embedded system safety and security, and automotive software. He received the 2018 IEEE TCSC Early Career Researcher Award. He is an IEEE senior member and ACM senior member.

Yawen Zhang is currently pursuing a master's degree in electronic information with Hunan University, China. She is a member of the Key Laboratory for Embedded and Network Computing of Hunan Province, China. Her current research interests include embedded and cyber-physical systems, parallel and distributed systems, and functional safety of automotive embedded systems.

Renfa Li is a Professor of computer science and electronic engineering with Hunan University, Changsha, China. He is the Director of the Key Laboratory for Embedded and Network Computing of Hunan Province, China. He is also an expert committee member of the National Supercomputing Center, Changsha. His major interests include computer architectures, embedded computing systems, cyber-physical systems, and Internet of things. Dr. Li is a member of the council of the China Computer Federation and a senior member of the Association for Computing Machinery.

Kenli Li is currently the dean and a full professor of computer science and technology with Hunan University and the director of the National Supercomputing Center in Changsha. His major research areas include parallel and distributed computing, edge computing, high-performance computing, grid, and cloud computing. He has published over 160 research papers in international conferences and journals. He is also a senior member of the IEEE.

Keqin Li is a SUNY Distinguished Professor of computer science with the State University of New York. He is also a National Distinguished Professor with Hunan University, China. His current research interests include cloud computing, fog computing and mobile edge computing, energy-efficient computing and communication, embedded systems, and cyber-physical systems. He is among the world's top five most influential scientists in parallel and distributed computing based on a composite indicator of the Scopus citation database. He is an IEEE Fellow.

Functional Safety for Embedded Systems

Guoqi Xie
Yawen Zhang
Renfa Li
Kenli Li
Keqin Li

CRC Press
Taylor & Francis Group
Boca Raton London New York

CRC Press is an imprint of the
Taylor & Francis Group, an **informa** business

Designed cover image: Guoqi Xie, Yawen Zhang, Renfa Li, Kenli Li, Keqin Li

Supported by the Key Projects of National Natural Science Foundation of China (NSFC) under the Grants 61932010 and 62133014.

First edition published 2023
by CRC Press
6000 Broken Sound Parkway NW, Suite 300, Boca Raton, FL 33487-2742

and by CRC Press
4 Park Square, Milton Park, Abingdon, Oxon, OX14 4RN

CRC Press is an imprint of Taylor & Francis Group, LLC

ISBN: 978-1-032-48936-0 (hbk)
ISBN: 978-1-032-48938-4 (pbk)
ISBN: 978-1-003-39151-7 (ebk)

DOI: 10.1201/9781003391517

Typeset in Latin Modern font
by KnowledgeWorks Global Ltd.

Publisher's note: This book has been prepared from camera-ready copy provided by the authors.

Contents

Section II SAFETY-AWARE COST OPTIMIZATION

Foreword

Embedded systems are widely used in many consumer electronics, entertainment devices, home appliances, industrial equipment, medical instruments, military weapons, and research facilities. They are extensively used in application areas such as aerospace control systems, the automobile industry, banking and finance, robotic systems, security and telecommunication, and traffic control.

Modern automobiles are typical safety-critical embedded systems, and development of self-driving systems is one of the hottest research areas in recent years. Meanwhile, the continuous advancement of embedded systems brings new functional safety requirements and design challenges. In recent years, some organizations have issued individual functional safety standards related to embedded systems, and the relevant functional safety research is gradually applied to practical applications. However, the functional safety assurance for embedded systems is a complex process, especially for parallel applications in distributed environments. There is an urgent need to design functional safety assurance techniques from the perspective of computing technology, thereby coping with respective characteristics and challenges of distributed embedded systems. The publication of this book satisfies this need in a timely manner.

The book introduces the functional safety standards related to embedded systems. It presents the design methods of functional safety assurance (including functional safety verification, enhancement, and validation), safety-aware hardware cost optimization, and safety-aware development cost optimization for embedded systems. The book combines the practical example of automotive embedded systems with the proposed functional safety design methods.

The book proposes various algorithms about functional safety assurance and safety-aware cost optimization for parallel applications of embedded systems. The book is rich in content and detailed in diagrams. A unique and effective feature of the book is to use appropriate motivational examples to clearly explain each proposed algorithm for the purpose of easier understanding. This book contains not only the basic knowledge and information, but also the latest research progress on the theory and methods of functional safety for embedded systems.

This book is the joint effort and endeavor of five scholars who have published very extensively in the fields of embedded computing, high-performance computing, embedded systems, and cyber-physical systems in the past few years. They are undoubtedly leading experts in the fields of embedded computing and high-performance computing. Their distinction and dedication make the book an important addition to the research community. The book is truly a significant contribution to the field of functional safety for embedded systems.

Finally, I would like to congratulate the authors for their solid work, and I look forward to seeing the book published.

Weimin Zheng
Member of the Chinese Academy of Engineering
Tsinghua University
Beijing, China

Preface

MOTIVATION OF THE BOOK

Ensuring functional safety is always a precondition in the realization of various embedded systems. However, the functional safety design of the system is challenged by multiple factors. Taking the automotive embedded system as an example, the complexity of the new generation automotive electrical and electronic (E/E) architecture, the continuous release and update of automotive functional safety standards, the publishing of a new AUTOSAR adaptive platform standard, and the increase in different kinds of costs bring challenges to functional safety design. Automotive systems are safety-critical embedded systems; consequences will be serious if functional safety cannot be guaranteed. Therefore, automobile manufacturers attach great importance to functional safety. In addition, the automobile industry is a cost-sensitive industry, so it is necessary to optimize costs while ensuring safety. This book uses the automotive embedded system as an example to introduce functional safety assurance and safety-aware cost optimization. The functional safety assurance integrates safety verification, enhancement, and validation. The safety-aware cost optimization divides cost types in terms of the essential differences of various costs in system design. The motivation of this book is to provide our recent research results on the aforementioned topics in recent years.

SUMMARY OF CONTENTS

Chapter 1 introduces functional safety for embedded systems. Most embedded systems are safety-critical systems; they must meet reliability and response time requirements simultaneously. This chapter takes the automotive embedded system as an example to introduce the functional design methods for ensuring functional safety, including functional safety verification, enhancement, and validation. The automotive industry is a cost-sensitive industry, and safety-aware cost optimization is a beneficial supplement to improve system design. Therefore, this chapter analyzes the necessities and challenges of hardware cost optimization and development cost optimization. Finally, this chapter lists the outline of this book.

Chapter 2 proposes a fast functional safety verification (FFSV) technique for parallel applications of embedded systems. First, this chapter presents the FFSV1 method to find the solution with the minimum response time under the reliability requirement. Second, this chapter presents the FFSV2 method to find the solution with the maximum reliability under the response time requirement. Finally, this chapter combines FFSV1 and FFSV2 to create the union FFSV (UFFSV). UFFSV is a fast heuristic method, and it can shorten the application's development lifecycle.

Chapter 3 studies functional safety enhancement techniques for parallel applications of embedded systems. This chapter presents forward safety enhancement (FFSE), repeated backward functional safety enhancement (RBFSE), and repeated FSE(RFFSE) algorithms to enhance the reliability values for a parallel application on automotive embedded systems. Considering that RBSE and RFSE could be invoked repeatedly until reaching a stable safety value, we propose the stable stopping-based safety enhancement (SSFSE) approach by combining the above algorithms. SSSE enhances the safety by using a stable stopping approach on the basis of the forward-and-backward recovery through primary-backup repetition.

Chapter 4 focuses on functional safety validation for parallel applications of embedded systems. This chapter proposes two effective reliability validation approaches, geometric mean-based non-fault-tolerant reliability pre-assignment (GMFRA), and geometric mean-based fault-tolerant reliability pre-assignment(GMFRA), for an automotive application based on geometric mean. These two approaches are used for the mechanisms of non-fault-tolerance and fault-tolerance, respectively.

Chapter 5 designs two hardware cost optimization methods. The first method proposes the progressive hardware cost optimization (PHCO), enhanced PHCO (EPHCO), and simplified EPHCO (SEPHCO) algorithms step by step for a distributed application while ensuring the functional safety requirement. The second approach proposes the cost-effectiveness-driven hardware cost optimization algorithm (CEHCO) for a distributed application while meeting the functional safety requirement.

Chapter 6 solves the problem of development cost optimization for an end-to-end embedded system function under ensuring the functional safety requirements based on the automotive safety integrity level (ASIL) decomposition defined in ISO 26262. First, this chapter proposes two heuristic algorithms, reliability calculation of scheme (RCS) and minimum development cost with reliability requirement (MDCRR) for parallel applications on distributed embedded systems. Second, this chapter presents FRA and DRA algorithms considering reliability and real-time requirements for real-time parallel applications on distributed embedded systems.

Chapter 7 summarizes the book and mentions future research.

AUDIENCE AND READERSHIP

This book should be a useful reference for researchers, engineers, and practitioners interested in embedded systems, Cyber-Physical Systems (CPSs), and functional safety of automotive embedded systems. This book can be used as a supplement to the advanced undergraduate or graduate courses of embedded computing, distributed computing, and CPSs in computer science, computing engineering, and electrical engineering. By reading this book, postgraduates and doctoral students will be familiar with the functional safety attributes of embedded systems, learn functional safety assurance and cost optimization algorithms, and find inspiration for their own research.

ACKNOWLEDGMENTS

This book was supported by the Key Projects of National Natural Science Foundation of China (NSFC) under the Grants 61932010 and 62133014, and the Outstanding Youth Fund of the Natural Science Foundation of Hunan Province under the Grant 2022JJ10021. The Grant 61932010 is titled "modeling theory and system design for safety- and security-critical automotive cyber-physical systems", the Grant 62133014 is titled "intelligent interconnection of things and integrated safety and security in industrial cyber-physical systems", and the Grant 2022JJ10021 is titled "real-time systems". Part of the research in this book was done in conjunction with researchers Gang Zeng, Hao Peng, Jia Zhou, Jinlin Song, Na Yuan, Shiyan Hu, Wei Wu, Wenhong Ma, Yan Liu, Yanwen Li, Yong Xie, Yuekun Chen, Yunbo Han, and Zhetao Li, and we are grateful for their contributions. The authors would like to express their gratitude to Professor Weimin Zheng, a member of the Chinese Academy of Engineering, for writing the Foreword of this book. We also thank Ms. Joy Luo of Taylor & Francis Group for her efforts and support in helping to publish this book.

Contributors

Guoqi Xie

Key Laboratory for Embedded and
Cyber-Physical Systems of Hunan
Province, College of Computer Science
and Electronic Engineering, Hunan
University
Changsha, Hunan, China

Yawen Zhang

Key Laboratory for Embedded and
Cyber-Physical Systems of Hunan
Province, College of Computer Science
and Electronic Engineering, Hunan
University
Changsha, Hunan, China

Renfa Li
Key Laboratory for Embedded and
Cyber-Physical Systems of Hunan
Province, College of Computer Science
and Electronic Engineering, Hunan
University
Changsha, Hunan, China

Kenli Li
College of Computer Science and Electronic
Engineering, Hunan University
Changsha, Hunan, China

Keqin Li
Department of Computer Science, State
University of New York
New Paltz, NY, USA

CHAPTER 1

Introduction

I NDUSTRY 4.0 strategy plans to extend embedded systems with information and communication technologies (ICT) based on automation, resulting in massively heterogeneous distributed embedded systems to complex parallel applications. An important safety issue needs to be addressed in embedded systems. According to the latest global estimates, taking into account the "Industry 4.0" strategic plan and the entire safety industry, the annual cost of workplace accidents amounts to 476 billion Euros. It is critical to ensure that embedded applications operate safely without causing personal injury and health damage. As advanced heterogeneous distributed systems, cyber-physical systems (CPS) further enhance the existing embedded systems. Specifically, Automotive CPS (ACPS) is a CPS applied to embedded areas. ACPS is the safety-critical embedded system, which has the attributes of high accuracy of time calculation, predictable time behavior, and strict deadlines. There are a large number of parallel applications with precedence constraints in those heterogeneous distributed embedded systems which can be described by a directed acyclic graph (DAG) at a high level. How to efficiently ensure functional safety for parallel applications of embedded systems is an important research direction and worth studying. Moreover, the automotive industry is a cost-sensitive industry, so it is necessary to reduce costs while ensuring functional safety of embedded systems.

1.1 AUTOMOTIVE EMBEDDED SYSTEMS

Automotive systems are safety-critical embedded systems, such as Autonomous Emergency Braking (AEB), Adaptive Cruise Control (ACC), and Automated Parking System (APS), must meet the functional safety requirements. It is necessary to reduce the risk of safety hazards at the beginning of the development lifecycle to further avoid injury or accidents. Compared with traditional safety-critical embedded systems, the complexity of automotive embedded systems is higher [55]. This book takes the automotive embedded system as an example to study the functional safety for embedded systems. A typical automotive electrical and electronic (E/E) architecture is shown in Fig. 1.1, where more than four or five Controller Area Network (CAN) buses are integrated by a central gateway and several Electronic Control Units (ECUs) are mounted on each CAN bus.

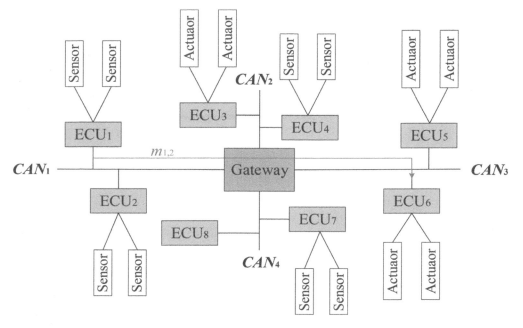

Figure 1.1: The architecture of automotive embedded systems.

1.2 FUNCTIONAL SAFETY

1.2.1 Functional Safety Standard

The safety issue is an important issue that needs to be addressed in embedded systems. According to new global estimates, the cost of workplace accidents and injuries in the context of Industry 4.0 and the safety industry as a whole reaches 476 billion Euros per year [3]. It is critical to ensure that the safety operation of parallel applications of embedded systems does not cause personal injury and personal health damage. Safety refers to the freedom from unacceptable risk of personal injury or personal health damage directly or indirectly due to damage to property or the environment [26]. Safety-critical parallel applications should operate in compliance with relevant functional safety standards. Safety issues are widely present in automotive electronics, industrial safety components, factory automation, process industries, nuclear power instrumentation control, railroad signaling, smart grid, and many other safety-related areas [28]. Functional safety focuses on the behavior after system failure, rather than the original function or performance of the system. Fig. 1.2 lists the main safety standards in the industry. Note that the standards in Fig. 1.2 include sections where functional safety is considered (i.e., only some sections, not all sections of these standards, describe functional safety).

IEC 61508, known as functional safety for electrical/electronic/programmable electronic safety-related systems (E/E/PES), is a basic functional safety standard applicable to various industries [26]. Functional safety standards for special industrial applications include IEC 61513 for nuclear power plants, IEC 61511 for the process industry, IEC 62061 for machinery, and IEC 61784 for industrial

Figure 1.2: Functional safety standards.

communications, among others. These standards are based on IEC 61508 and complement IEC 61508 in specific areas. These standards change will over time; for example, the latest version of IEC 61513 is IEC 61513:2011, which was published in 2011. ISO 26262 is a functional safety standard formulated for eight-seat passenger cars with a gross weight of no more than 3.5 tons and based on the attributes of safety related electronic and electrical systems. It was formulated based on IEC 61508, officially released on November 15, 2011, and updated in 2018.

1.2.2 Automotive Safety Integrity Level (ASIL) Determination

This subsection introduces the Automotive Safety Integrity Level (ASIL) as it relates to the safety assessment of parallel applications of embedded systems. ASIL is a risk classification system defined by the ISO 26262 standard for the functional safety of road vehicles. ASIL is established by performing a hazard analysis and risk assessment (HARA) of potential hazards by assessing the severity, exposure, and controllability of automotive operating scenarios. ASIL is determined by a combination of three factors: severity, exposure, and controllability [28]. Table 1.1 shows the classifications of severity, exposure, and controllability in ISO 26262. ISO 26262 identifies five ASILs:

Table 1.1: Classification of severity, exposure, and controllability levels in ISO 26262 [28].

	Severity		Exposure		Controllability
S0	No injuries	E0	Incredibly unlikely	C0	Controllable in general
S1	Light to moderate injuries	E1	Very low probability	C1	Simply controllable
S2	Severe to life-threatening injuries	E2	Low probability	C2	Normally controllable
S3	Life-threatening to fatal injuries	E3	Medium probability	C3	Difficult to control or uncontrollable
		E4	High probability		

Quality Management (QM), ASIL A, ASIL B, ASIL C, and ASIL D [28, 30]. QM means that no additional safety measures are required as long as the standard quality management process is followed in software development. ASIL A represents the lowest level of automotive hazards, while ASIL D represents the highest level of automotive hazards. Airbags, anti-lock braking systems, and power steering systems must achieve ASIL D, the most stringent level applied for safety and security, as their failure poses the highest risk. The lowest level of the functional safety rating range, such as components such as rear lights, only needs to achieve ASIL A. Headlights and brake lights are typically ASIL B, while automotive cruise control is typically ASIL C.

Severity refers to the degree of harm suffered by relevant persons when a hazardous event occurs. The relevant people include both the driver and passengers in the car, as well as other traffic participants, such as pedestrians on the road. ISO 26262 divides the severity into four levels: S0, S1, S2, and S3, and the severity increases as the number increases. Exposure refers to the exposure probability of the corresponding scenario of a hazard event. ISO 26262 divides exposure into five levels: E0, E1, E2, E3, and E4. Controllability refers to the degree to which the driver or other traffic participants can control the hazard when a hazard event occurs, or avoid specified harm or damage. ISO 26262 classifies controllability into four levels: C0, C1, C2, and C3. It should be noted that, unlike severity and exposure, the larger the value of C, the lower the controllability. ASIL combines the severity, exposure, and controllability of the aforementioned levels to obtain ASIL A, B, C, and D as shown in Table 1.2 [30].

Each ASIL is determined by a combination of severity, exposure, and controllability values. There are minor differences in ASIL determinations between Version 1 and Version 2 of the ISO 26262 standard. In Version 1, the combination of S3, E1, and C3 is ASIL A. However, in Version 2, if several unlikely scenarios are combined, the combination may be QM and the change may result in a lower probability of exposure than E1. QM is not ASIL but can be specified in HARA [28, 30]. Severity has been determined after HARA and cannot be changed. Controllability is a fixed value in the design phase. Therefore, a feasible measure to improve the safety of

Table 1.2: ASIL determination in ISO 26262 [28].

Severity	Exposure	Controllability		
		C1	C2	C3
S1	E1	QM	QM	QM
	E2	QM	QM	QM
	E3	QM	QM	A
	E4	QM	A	B
S2	E1	QM	QM	QM
	E2	QM	QM	A
	E3	QM	A	B
	E4	A	B	C
S3	E1	QM	QM	A/QM
	E2	QM	A	B
	E3	A	B	C
	E4	B	C	D

parallel applications of automotive embedded systems is to reduce the exposure of the application according to Table 1.2.

1.3   CHALLENGES OF FUNCTIONAL SAFETY DESIGN

In recent years, the latest advances of functional safety have been widely discussed. However, functional safety design methodology still faces the following severe challenges.

(1) The first edition of the functional safety standard ISO 26262, released in 2011, provided basic guidelines for the functional safety-based development process for automotive embedded systems. The second edition of ISO 26262 further introduced additional changes and additions to the safety development of automotive embedded systems, including functional safety management, support processes, and semiconductor guidelines. In particular, the concept of fault tolerance, the ability to deliver a specified application in the presence of one or more faults, was formally introduced. In recent years, some progress has been made in the design methodology for functional safety assurance, including functional safety verification, functional safety enhancement, and functional safety validation.

(2) The cost associated with automotive embedded systems has accounted for 30%–40% of the total automotive cost, with a growing trend. Cost has become one of the most important indicators of the design quality of the automotive embedded system. Given the cost sensitivity of the automotive industry, there is a need to optimize cost to improve design quality while satisfying functional safety requirements. However, the cost of an automotive embedded system is influenced by a number of complicated factors, such as hardware cost, development cost, etc.

A combination of the above challenges, automotive functional safety assurance has been challenged by many factors, such as the complexity of the new generation

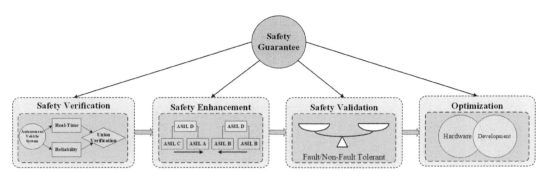

Figure 1.3: The structure of automotive embedded systems.

of automotive E/E architecture, the continuous release and update of the automotive functional safety standard, and the new release of the automotive open system architecture (AUTOSAR) adaptive platform standards, as well as different types of cost increases. These challenges demonstrate that while ISO 26262 provides many safety design specifications to guide the design and development of automotive embedded systems, it is not enough to design and develop automotive embedded systems according to ISO 26262.

1.4 STRUCTURE OF THE RESEARCH

To cope with the above challenges, we have studied automotive functional safety design methodologies, which are described through functional safety assurance and safety-aware cost optimization, respectively: (1) functional safety verification; (2) functional safety enhancement; (3) functional safety validation; (4) hardware cost optimization; and (5)development cost optimization. The structure of the research is shown in Fig. 1.3. The functional safety assurance method realizes a set of safe and efficient assurance systems that integrates functional safety verification, functional safety enhancement, and functional safety validation. The security-aware cost optimization method classifies cost types based on the essential differences of various costs in system design, and establishes reasonable cost models based on different costs. Functional safety verification is timely remediation of failed functional safety analysis results, while functional safety validation is the feedback and consolidation of functional safety verification and enhancement. Safety-aware cost optimization is a useful complement to improve system design. The above aspects support each other and jointly build the architecture of functional safety design methodologies for automotive embedded systems.

1.5 FUNCTIONAL SAFETY ASSURANCE

1.5.1 Functional Safety Verification

Functional safety verification can be performed after analyzing the time and reliability of the automotive application separately. Functional safety verification is to judge whether the application achieves its functional safety requirements [28, 30].

Functional Safety verification includes real-time verification and reliability verification of the system. If the tasks of systems cannot be finished within their deadline, it may cause automotive systems to fail. Therefore, it is necessary to regard time as one of the core attributes of functional safety. However, one of the difficulties in verification is that real-time and reliability requirements conflicts with each other, and the relationship between time and reliability presents a Pareto curve [20, 63]. How to conduct joint verification and increase the verification acceptance rate is a challenge.

1.5.2 Functional Safety Enhancement

Functional safety is a concern throughout the lifecycle of automotive systems development, so the maximum possible safety values should be known in the early design phase to help control risk during actual development. Functional safety enhancements, which imply increasing safety values (e.g., reducing exposure levels by increasing reliability values), are desirable for risk control. Considering that real-time requirements cannot be changed, increasing reliability is a common practice to enhance safety. ISO 26262 does not provide the concept of functional safety enhancement, but it is actually consistent with the safety mechanisms described in ISO 26262 for error handling (refer to Section 7.4.12, Part 6 of the second edition of ISO 26262) [28, 30]). ISO 26262 provides a variety of functional safety enhancement techniques, including static recovery mechanisms, deactivation, graceful degradation, homogeneous redundancy, diverse redundancy, corrective codes for data, and access permission management [30]. Among these techniques, static recovery mechanisms have recently been used for functional safety enhancement in parallel applications of automotive embedded systems. Static recovery mechanisms include recovery blocks, forward recovery, backward recovery, and recovery by repetition.

1.5.3 Functional Safety Validation

Functional safety validation means that the safety requirement is adequate and has been achieved with a sufficient level of integrity based on examination and tests [28, 30]. Safety validation is a comprehensive and sufficient judgment of whether the system meets the safety requirement, and is divided into non-fault-tolerant validation methods and fault-tolerant validation methods. Safety validation is a complete machine confirmation process. In order to make the comprehensive value of reliability between components more centralized, it is necessary to improve the accuracy of safety validation in a more balanced way and reduce the impact on time. In recent years, automotive functional safety validation methods have been deeply studied by introducing a reliability preassignment strategy.

1.6 SAFETY-AWARE COST OPTIMIZATION

The automotive industry is cost sensitive because automobiles are mass-produced industrial products. An increasing cost is an urgent problem to be solved because automakers and component suppliers have to pursue high profits in the fierce market competition [19, 56]. After achieving functional safety assurance for automotive applications, it is necessary to optimize the cost while ensuring functional safety requirements to improve the design quality. However, the cost types of automotive embedded systems are complex and the costs can be broadly divided into two categories: (1) hardware cost is the prices of hardware devices such as ECUs; (2) development cost is the amount of programmer's labor in developing the application.

1.6.1 Hardware Cost Optimization

Reducing hardware costs can result in significant savings in industrial production costs and high profits. While distributed architectures greatly reduce the hardware cost of a harness, the number of ECUs required for application execution still incurs significant hardware costs. In addition, as the size of the application continues to grow, the required process or hardware costs increase accordingly. Currently, the unit price of ECUs are from $5 to $100 [27]. In fact, some unnecessary ECUs can be removed, as long as such operations do not interfere with the correct execution of the application. Therefore, reducing the number of ECUs is crucial to reduce hardware cost [22].

1.6.2 Development Cost Optimization

The development process for safety-critical automotive applications is a fully structured and systematic process because of the compliance with ISO 26262 [56]. However, the standard increases development cost by adding additional processes and complexity as well as hardware/software redundant solutions in automotive application development and testing [56]. The development cost can be reduced by 25%–100% when decomposing a high ASIL task to a low ASIL one [56]. Therefore, ASIL decomposition is commonly applied for development cost optimization.

1.7 OUTLINE OF THE BOOK

The outline of this book is below.

In Chapter 2, we propose the fast functional safety verification (FFSV) for the parallel application of automotive embedded system containing two algorithms, which are FFSV1 and FFSV2 algorithms. FFSV1 algorithm is dedicated to minimizing response time under the reliability requirement, and FFSV2 algorithm is dedicated to maximizing reliability under the real-time requirement. Finally, FFSV1 and FFSV2 are combined to form the union FFSV (UFFSV). UFFSV can obtain a high acceptance rate than FFSV1 and FFSV2.

In Chapter 3, we present the reliability enhancement technique called stable stopping-based functional safety enhancement (SSFSE) for the parallel application of an automotive embedded system. SSFSE consists of existing HEFT and backward functional safety enhancement (BFSE) algorithms and pretested forward functional safety enhancement (FFSE), repeated BFSE (RBFSE), and repeated FFSE (RFFSE) algorithms. SSFSE enhances the safety by using a stable stopping approach on the basis of the forward-and-backward recovery through primary-backup repetition.

In Chapter 4, we first introduce the geometric mean-based non-fault tolerant reliability preassignment (GMNRA) method. GMNRA method is a non-fault tolerant method, which is difficult to validate the high reliability requirement due to the lacking of replication. Hence, fault tolerant methods should be adopted for a high reliability requirement. We develop the geometric mean-based fault tolerant reliability preassignment (GMFRA) method, which is a quantitative active replication strategy.

In Chapter 5, we design the progressive hardware cost optimization (PHCO), enhanced PHCO (EPHCO), and simplified EPHCO(SEPHCO) algorithms step by step for a parallel application while ensuring the functional safety requirements. The core idea of EPCHO is iteratively removing the ECUs (i.e., open-to-close) until the functional safety requirements cannot be met. We further propose cost-effectiveness-driven hardware cost optimization (CEHCO) for a parallel application while ensuring the functional safety requirements. CEHCO combines CEHCO1 (closed-to-opened) algorithm and CEHCO2 (opened-to-closed) algorithm, thereby realizing powerful cost optimization and superior time efficiency simultaneously.

In Chapter 6, we solve the development cost optimization problem for parallel applications of embedded systems. First, we devise the minimizing development cost under reliability requirement (MDCRR) algorithm based on ASIL decomposition to optimize the development cost of a parallel automotive application while ensuring its time requirement. The automotive embedded system is a real-time embedded system that needs simultaneously consider the real-time and reliability requirements. Therefore, we present the dual requirement assurance (DRA) algorithm simultaneously considering the real-time requirement and reliability requirement. DRA is a fast algorithm for embedded system functional safety risk assessment, which decides the feasibility of the development cost optimization.

In Chapter 7, we conclude the book and give some future research directions.

1.8 CONCLUDING REMARKS

In this chapter, we introduce functional safety for embedded systems. Safety-critical embedded systems must meet reliability and real-time requirements. We take the automotive embedded system as an example to analyze the design methods that can assurance functional safety, including functional safety verification, enhancement, and validation. In addition, the automotive industry is a cost-sensitive industry, and safety-aware cost optimization is a useful supplement to improve system design. We analyze the necessity and challenges of hardware cost optimization and development cost optimization. Finally, we list the outline of this book.

I

FUNCTIONAL SAFETY ASSURANCE

CHAPTER 2

Functional Safety Verification

B OTH reliability and response time are important functional safety attributes that must be simultaneously assured by learning from the automotive functional safety standard ISO 26262. Functional safety verification is the process of checking that an application complies with a set of safety design specifications and meets the requirements. Introducing verification in the early design phase not only complies with the latest automotive functional safety standards, but also avoids unnecessary design work or reduces the design burden in the later design optimization phase. This chapter presents fast functional safety verification (FFSV) algorithms for the parallel application of an automotive embedded system, including 1) FFSV1; 2) FFSV2; and 3) union FFSV (UFFSV). FFSV1 is committed to minimizing response time under the reliability requirement. FFSV2 is committed to maximizing reliability under the real-time requirement. Combining FFSV1 and FFSV2 to create UFFSV, which can obtain a higher acceptance rate than the current method. Finally, we do experiments with real-life and synthetic parallel applications of automotive embedded systems for functional safety verification to validate the performance and efficiency of UFFSV.

2.1 INTRODUCTION

In order to enhance the safety of automotive embedded systems, many active and passive safety applications, such as antilock braking system, brake-by-wire, and adaptive cruise control, have been designed [65]. Functional safety refers to the absence of unreasonable risk caused by systematic failures and random hardware failures in ISO 26262 [28]. Safety usually refers to satisfying the reliability requirement (i.e., reliability assurance, and reliability constraint) and real-time requirement (i.e., response time requirement, and timing constraint) of an application. References [19, 22, 56] have studied safety optimization for safety-critical applications of the automotive embedded system but these research only focused on either meeting the response time or reliability requirement rather than functional safety requirements. In the ISO 26262 standard, response time and reliability are important functional safety attributes; their requirements must be simultaneously satisfied for automotive functional safety

DOI: 10.1201/9781003391517-2

[28]. Therefore, we consider verifying the functional safety requirements during early design phase of automotive applications. On the one hand, introducing verification during the early design phase tallies with the latest automotive functional safety standard; on the other hand, it also avoids unnecessary design effort or reduces the development burden of the later design phases.

The problem is that real-time requirement and reliability requirement may not be satisfied simultaneously in practice because increasing reliability intuitively increases the response time of a parallel application [20, 63]. The exposure is defined in ISO 26262 to represent the relative expected frequency of the operational conditions, in which dangerous events may happen and cause hazards and injuries [28]. That is, reliability is just the inverse expression of exposure. Minimizing response time and minimizing exposure (i.e. maximizing reliability) conflict with each other; thus, verifying functional safety is a bi-objective optimization problem. In Fig. 2.1, each point of x^1-x^7 represents a solution of a bi-objective minimization problem [20, 63]. The points x^1, x^2, x^3, x^4, and x^5 are Pareto optima; the points x^1 and x^5 are weak optima, whereas the points x^2, x^3, and x^4 are strong optima. The set of all Pareto optima is called Pareto curve.

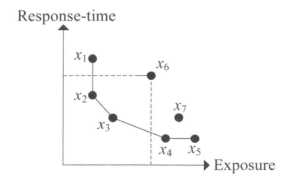

Figure 2.1: Pareto curve for a bi-criteria between response time and exposure [20, 63].

At present, the high-end automotive system is composed of at least 70 heterogeneous ECUs. It is expected that the number of ECUs in the automotive system will further increase in the future [65, 72]. Considering the cost sensitivity of the automotive industry, it is very important to shorten the development cycle of automotive applications to reduce development costs. Therefore, a fast functional safety verification (FFSV) technique with low time complexity should be presented from the perspective of cost control.

In this chapter, we discuss functional safety verification for the parallel application of automotive embedded systems and devise two heuristic verification algorithms to implement union fast functional safety verification (UFFSV). The details are summarized as follows.

(1) We present the FFSV1 algorithm to address the problem of response time minimization under reliability requirement. The problem is divided into two sub-problems, (1) satisfying reliability requirement; (2) minimizing response time.

The solution to the first sub-problem is to transfer the reliability requirement of the parallel application to each task, and the solution to the second sub-problem is to assign each task to the ECU with the minimum earliest finish time (EFT) under satisfying the reliability requirement of the parallel application.

(2) We present the FFSV2 algorithm to address the problem of reliability maximization under real-time requirement. The problem is divided into two sub-problems, (1) satisfying real-time requirement; (2) maximizing reliability. The solution to the first sub-problem is to transfer the real-time requirement of the parallel application to each task, and the solution to the other one is to migrate partial tasks to the ECUs with maximum reliability values without violating the real-time requirement of the parallel application.

(3) We combine FFSV1 and FFSV2 to form UFFSV. UFFSV can obtain a higher acceptance rate than the current method..

2.2 RELATED WORK

As stated in ISO 26262, random hardware failures (i.e., transient failures in most studies) occur unpredictably during the lifecycle of a hardware element, but they follow a probability distribution [28]. Reference [51] presented a widely accepted reliability model, in which the transient failure of each hardware follows a constant-parameter Poisson law [20, 61]. An automotive system is a safety-critical embedded system, both reliability and real-time requirements must be simultaneously satisfied by learning from automotive functional safety standards.

Minimizing response time and maximizing reliability are conflicting in scheduling a parallel application, and optimizing them is a bi-objective optimization problem [20]. Reference [20] devised a bi-objective scheduling heuristic (BSH) to generate an approximate Pareto curve of non-dominated solutions, among which the designers can verify the functional safety by finding the points that meet the real-time and reliability requirements simultaneously. However, the time complexity of BSH is as high as $O(|N| \times 2^{|U|})$, where $|N|$ and $|U|$ are the number of tasks and electronic control units (ECUs), respectively.

Reference [74] proposed a shared recovery-based frequency allocation approach to reduce energy consumption with a reliability requirement and a real-time requirement for a parallel application on a single ECU. However, multiple ECUs have been used in most embedded systems, such as automotive embedded systems. Reference [61] addressed the issue of minimizing the resource consumption cost by satisfying the reliability requirement without using fault-tolerance on heterogeneous embedded systems. Reference [63] solved the problem of minimizing the redundancy by meeting the reliability requirement using fault-tolerance on heterogeneous service-oriented systems. Given the limited resources of embedded systems, fault-tolerance may be unsuitable [61].

2.3 MODELS AND PRELIMINARIES

Tables 2.1 and 2.2 list the abbreviations and notations that are used in this chapter.

Table 2.1: Abbreviations in this chapter.

Abbreviation	Definition
LB	Lower bound
WCET	Worst Case Execution Time
WCRT	Worst Case Response Time
ECU	Electronic Control Unit
CAN	Controller Area Network
DAG	Directed Acyclic Graph
ET	End Time
ST	Start Time
MET	Maximum Execution Time
EST	Earliest Start Time
EFT	Earliest Finish Time
LFT	Latest Finish Time
AST	Actual Start Time
AFT	Actual Finish Time
HEFT	Heterogeneous Earliest Finish Time
FFSV	Fast Functional Safety Verification

2.3.1 System Model

This chapter discusses a typical integration architecture where several electronic control units (ECUs) are mounted on the controller area network (CAN) bus as shown in Fig. 2.2 [56]. CAN is a half duplex, static priority, and non-preemptive scheduling communication bus; thus, it is ideally suited for distributed automotive embedded systems [41, 64, 71, 72]. ECUs can connect to several sensors or several actuators because physical processes are composed of many parallel processes [65]. Partial ECUs can release the function point by receiving collected data from sensors, and other partial ECUs can complete the function point by sending the performing action to the actuators [65]. One task sends messages to all its successor tasks when it has finished executing on one ECU. Tasks may be executed on the different ECUs. For instance, task n_1 is executed on ECU u_1. Then n_1 sends a message $m_{1,2}$ to its successor task n_2 executed on ECU u_4. $U = \{u_1, u_2, ..., u_{|U|}\}$ represents a set of heterogeneous ECUs, where $|U|$ represents the size of set U. We use $|X|$ to denote the size of the set X in this book.

A parallel application of an embedded system is represented by a DAG $G=(N, W, M, C)$ [58, 61, 65].

(1) N represents a set of tasks in the parallel application G, and n_i represents the ith task of $G(n_i \in N)$. The set of the immediate predecessor tasks of n_i is

Table 2.2: Notations in this chapter

Notations	Definition
n_i	A computing task in a parallel application
$m_{i,j}$	A CAN message from tasks n_i to n_j
$w_{i,k}$	WCET of the task n_i on the ECU u_k
$c_{i,j}$	WCRT between the tasks n_i and n_j
$rank_u(n_i)$	Upward rank value of the task n_i
$pred(n_i)$	Set of n_i's immediate predecessor tasks
$succ(n_i)$	Set of n_i's immediate successor tasks
$u_{pr(i)}$	Assigned ECU of the task n_i
$|X|$	Size of the set X
λ_k	Failure rate of the ECU u_k
$n_{seq(y)}$	yth assigned task of the parallel application
$R(n_i, u_k)$	Reliability of the task n_i on the ECU u_k
$R(n_i)$	Reliability of the task n_i
$R_{req}(n_i)$	Reliability requirement of the task n_i
$R_{max}(n_i)$	Maximum reliability of the task n_i
$R(G)$	Reliability of the parallel application G
$R_{max}(G)$	Maximum Reliability of the parallel application G
$R_{req}(G)$	Reliability requirement of the parallel application G
$RR(G)$	Reliability ratio of the parallel application G
$R_{rrp}(G)$	Rate-based reliability preassignment of the parallel application G
$R_{rrp}(n_i)$	Rate-based reliability preassignment of the task n_i
$LB(G)$	Lower bound of the parallel application G
$RT(G)$	Response time of the parallel application G
$EST(n_i, u_k)$	Earliest start time of the task n_i on the ECU u_k
$EFT(n_i, u_k)$	Earliest finish time of the task n_i on the ECU u_k
$AST(n_i)$	Actal start time of the task n_i
$AFT(n_i)$	Actual finish time of the task n_i
$RT_{req}(n_i)$	Real-time requirement of the task n_i
$RT_{req}(G)$	Real-time requirement of the parallel application G

denoted by pred(n_i), and the set of the immediate successor tasks of n_i is denoted by succ(n_i). n_{entry} is used to represent the task with no predecessor task, and n_{exit} is used to represent the task with no successor task. In automotive embedded systems, the ECUs can accept data from various sensors and send the messages to multiple actuators, the parallel application may have multiple entry tasks or exit tasks. If the parallel application has multiple n_{entry} or n_{exit} tasks, a dummy entry or exit task with zero-weight dependencies is added to the G.

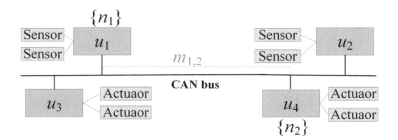

Figure 2.2: Integration architecture of automotive embedded systems.

(2) W is a $|N| \times |U|$ matrix, where $w_{i,k}$ represents the worst case execution time (WCET) of n_i executing on the ECU u_k. Due to the heterogeneity of ECUs, each n_i has different WCET values on different ECUs [23]. The WCET of one task is the maximum execution time among all possible real execution time values when the task is executed on a specific ECU with the maximum frequency. All the WCETs of the $n_i \in N$ are determined by the analysis method performed (i.e. WCET analysis [8]) in the analysis phase [65].

(3) The communication between tasks mapped to different ECUs is performed by passing messages on the bus. Hence, M is a set of messages, and each edge $m_{i,j} \in M$ represents the communication message from n_i to n_j. Accordingly, the worst case response time (WCRT) of $m_{i,j}$ is represented by $c_{i,j} \in C$ [65]. The WCRTs for all messages are also determined by performing analysis methods (i.e., WCRT analysis) [65].

Scheduling can be either non-preemptive(e.g., eCos) or preemptive (e.g., OS-EKTime) in automotive embedded systems [65]. In practice, many DAG-based parallel application scheduling algorithms generally use non-preemptive scheduling [58, 61, 65], we only consider non-preemptive scheduling for ECUs in this chapter. Of course, the approach of this chapter can also be applied to preemptive scheduling.

Fig. 2.3 shows a motivational parallel application with 10 tasks and 15 messages [58, 61]. The example shows 10 tasks running on three ECUs. The weight 9 of the edge between n_1 and n_4 represents the WCRT, denoted by $c_{1,4}$ if n_1 and n_4 are not assigned to the same ECU. Table 2.3 is the WCET matrix $|N| \times |U|$ of parallel application in Fig. 2.3. For instance, the weight 14 of n_1 and u_1 in Table 2.3 represents the WCET of n_1 on u_1, represented by $w_{1,1} = 14$. The same task n_i has different WCETs on different ECUs due to the heterogeneity of the ECUs. This chapter will illustrate the devised verification approaches with the motivational application. For the sake of simplicity, the units of all parameters are ignored in this example.

2.3.2 Reliability Model

The random hardware failures (i.e., transient failures) and the permanent failures are two major temporal types of failures [20, 61]. We only consider the random hardware failures of ECUs due to the ISO 26262 standard only combines the transient

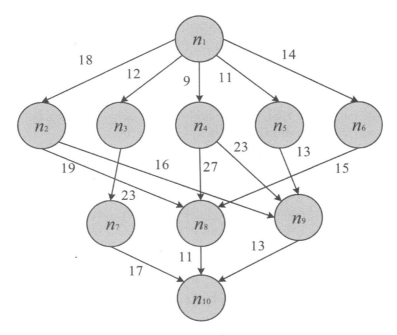

Figure 2.3: A motivational parallel application of the embedded system [58, 61].

Table 2.3: WCETs of tasks on different ECUs of the motivational parallel application [58, 61].

Task	u_1	u_2	u_3
n_1	14	16	9
n_2	13	19	18
n_3	11	13	19
n_4	13	8	17
n_5	12	13	10
n_6	13	16	9
n_7	7	15	11
n_8	5	11	14
n_9	18	12	20
n_{10}	21	7	16

failure and reliability together [28]. ISO 26262 stipulates that transient failure occurs unpredictably in the lifecycle of hardware, but follows the probability distribution [28]. Normally, transient failure for a task in a parallel application follows the Poisson distribution [20, 61]. The reliability of an event in unit time t is denoted by $R(t) = e^{-\lambda t}$, where λ is the constant failure per unit time (i.e., failure rate) for an ECU. We let λ_k represent the constant failure rate per time unit of the ECU u_k, and the reliability of n_i executed on u_k is denoted by

$$R(n_i, u_k) = e^{-\lambda_k w_{i,k}}. \tag{2.1}$$

The reliability of the parallel application is calculated by [20, 61]

$$R(G) = \prod_{n_i \in N} R(n_i, u_{\text{pr}(n_i)}), \qquad (2.2)$$

where $u_{\text{pr}(n_i)}$ represents the assigned ECU of n_i. As CAN bus has a high fault tolerance capacity, we only consider ECU failure and do not factor the communication failure into the problem (i.e., communication is assumed reliable).

2.3.3 Reliability Requirement Assessment

As the WCET of each task n_i on each ECU has been determined by the WCET analysis during the analysis phase, the maximum reliability value of task n_i can be obtained by traversing all the ECUs, and it is calculated by

$$R_{\text{max}}(n_i) = \max_{u_k \in U} R(n_i, u_k). \qquad (2.3)$$

Since the reliability of the parallel application G is the product of reliability values of all tasks (Eq. (2.2)), the maximum reliability value of the parallel application G is calculated by

$$R_{\text{max}}(G) = \prod_{n_i \in N} R_{\text{max}}(n_i). \qquad (2.4)$$

The reliability requirement $R_{\text{req}}(G)$ must be less than or equal to the maximum reliability value $R_{\text{max}}(G)$. Accordingly, $R_{\text{req}}(G)$ must satisfy the following constraint:

$$R_{\text{req}}(G) \leqslant R_{\text{max}}(G), \qquad (2.5)$$

otherwise, the reliability requirement assessment cannot be passed. For instance, we assume that the failure rates of ECUs u_1, u_2, and u_3 are $\lambda_1 = 0.0002$, $\lambda_2 = 0.0005$, and $\lambda_3 = 0.0009$, respectively. We can calculate that the maximum reliability value of the parallel application is $R_{\text{max}}(G) = 0.974335$. We set $R_{\text{req}}(G) = 0.96$, which can pass the assessment, because

$$0.96 \leqslant 0.974335$$

according to Eq. (2.5) of the reliability requirement assessment.

2.3.4 Real-Time Requirement Assessment

Task scheduling for multiple ECUs is an NP-hard problem [59]; thus, obtaining a minimum response time of a parallel application is also an NP-hard optimization problem [58]. Reference [58] proposed the heterogeneous earliest finish time (HEFT) algorithm for minimizing response time, and it is a widely accepted heuristic list scheduling algorithm and applied for the real-time requirement assessment of parallel applications [65]. The important steps of the HEFT algorithm are as follows:

(1) **Task prioritization.** HEFT uses the task's upward rank value ($rank_u$) (Eq. (2.6)) as the task priority criterion. Therefore, tasks are ordered in descending order of $rank_u$, with $rank_u$ calculated by

$$rank_u(n_i) = \overline{w_i} + \max_{n_j \in succ(n_i)} \{c_{i,j} + rank_u(n_j)\}, \tag{2.6}$$

where w_i represents the average WCET of task n_i.

Table 2.4 shows the $rank_u$ of all the tasks in Fig. 2.3. Note that n_i is ready to be assigned only when all predecessors of n_i have been assigned. We assume that n_i and n_j meet $rank_u(n_i) > rank_t extu(n_j)$. If there is no precedence constraint between n_i and n_j, n_i is not necessarily assigned in preference to n_j. Hence, the task assignment order in G is $n_1, n_3, n_4, n_2, n_5, n_6, n_9, n_7, n_8$, and n_{10}.

(2) **Response time minimization.** The earliest start time (EST) and the earliest finish time (EFT) need to be calculated. The attributes $EST(n_i, u_k)$ and $EFT(n_i, u_k)$ represent the EST and EFT, respectively, of task n_i on ECU u_k. $EFT(n_i, u_k)$ is considered as the task assignment criterion because it satisfies the local optimum of each task. The aforementioned attributes are calculated as follows:

$$\begin{cases} EST(n_{\text{entry}}, u_k) = 0, \\ EST(n_i, u_k) = \max \left\{ \begin{array}{l} avail[k], \\ \max\limits_{n_h \in pred(n_i)} \{AFT(n_h) + c'_{h,i}\} \end{array} \right\}; \end{cases} \tag{2.7}$$

and

$$EFT(n_i, u_k) = EST(n_i, u_k) + w_{i,k}. \tag{2.8}$$

$avail[k]$ is the earliest available time at which u_k is ready to perform the task. $AFT(n_h)$ is the actual finish time (AFT) of task n_h. $c'_{h,i}$ denotes the WCRT between n_h and n_i. If n_h and n_i are assigned to the same ECU, then $c'_{h,i} = 0$; otherwise, $c'_{h,i} = c_{h,i}$. By using an insertion-based scheduling strategy, n_i is assigned to the ECU with minimum EFT, where n_i can be inserted into the slack with minimum EFT.

The real-time requirement assessment processes are as follows.

(1) The response time obtained by HEFT represents the lower bound of the parallel application. The lower bound refers to the minimum response time of a parallel application generated by the HEFT algorithm, and is calculated as follows:

$$LB(G) = AFT(n_{exit}), \tag{2.9}$$

where n_{exit} represents the exit task as mentioned earlier. Fig. 2.4 shows the task mapping of lower bound calculation of the motivational application, where $LB(G) = 80$. Note that the arrows in Fig. 2.4 represent the generated communications between precedence constrained tasks.

Table 2.4: Upward rank values for tasks of the motivational parallel application.

Task	$rank_u(n_i)$
n_1	108
n_2	77
n_3	80
n_4	80
n_5	69
n_6	63.3
n_7	42.7
n_8	35.7
n_9	44.3
n_{10}	14.7

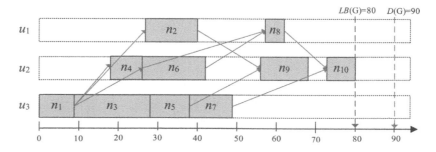

Figure 2.4: Task mapping for lower bound calculation.

(2) The known real-time requirement $RT_{\text{req}}(G)$ (i.e., deadline) is provided for the parallel application on the basis of the actual physical time requirement after hazard analysis and risk assessment (Fig. 2.4). $RT_{\text{req}}(G)$ must be larger than or equal to $LB(G)$. Therefore, $RT_{\text{req}}(G)$ must meet the following constraint:

$$LB(G) \leqslant RT_{\text{req}}(G). \tag{2.10}$$

Otherwise, the motivational parallel application does not pass the assessment of the real-time requirement. For the application G, we let the real-time requirement be $RT_{\text{req}}(G) = 90$, which passes the assessment, because

$$80 \leqslant 90$$

according to Eq. (2.10) of the reliability requirement assessment.

2.3.5 Problem Statement

The problem description of this chapter is to verify whether a solution exists for the following conditions:

$$RT(G) \leqslant RT_{\text{req}}(G), \tag{2.11}$$

and

$$R(G) \geqslant RT_{\text{req}}(G) \tag{2.12}$$

can be simultaneously satisfied. If the result is true, then the functional safety verification is passed and the later cost design optimization is feasible; otherwise, the functional safety verification is failed and the later cost design optimization is not feasible.

2.4 RESPONSE TIME MINIMIZATION UNDER RELIABILITY REQUIREMENT

This section proposes the first functional safety verification approach by minimizing the response time while satisfying the reliability requirement.

2.4.1 Satisfying Reliability Requirement

Reference [61] addressed the problem of resource consumption cost minimization while satisfying the reliability requirement by presenting the MRCRG algorithm, where the strategy of satisfying the reliability requirement is as follows. Assuming that the task to be assigned is $n_{s(y)}$, where $s(y)$ denotes the y-th assigned task (sequence number), $\{n_{s(1)}, n_{s(2)}, \dots, n_{s(y-1)}\}$ represents the task set where the tasks have been assigned, and $\{n_{s(y+1)}, n_{s(y+2)}, \dots, n_{s|N|}\}$ represents the task set where the tasks have not been assigned. The tasks are sorted according to the descending order of $rank_u$ values as done in the HEFT algorithm. To ensure that the reliability of the parallel application is met at each task assignment, we presuppose that each unassigned task in $\{n_{s(y+1)}, n_{s(y+2)}, \dots, n_{s|N|}\}$ is assigned to the ECU with the maximum reliability value. Hence, when assigning $n_{s(y)}$, the constraint of reliability of G is as follows:

$$R_{s(y)}(G) = \prod_{x=1}^{y-1} R(n_{s(x)}) \times R(n_{s(y)}) \times \prod_{z=y+1}^{|N|} R_{\max}(n_{s(z)}) \geqslant R_{\text{req}}(G). \tag{2.13}$$

Then, $R(n_{s(y)})$ must derive the following constraint:

$$R(n_{s(y)}) \geqslant \frac{R_{\text{req}}(G)}{\prod_{x=1}^{y-1} R(n_{s(x)}) \times \prod_{z=y+1}^{|N|} R_{\max}(n_{s(z)})}. \tag{2.14}$$

MRCRG sets the reliability requirement of the task $n_{s(y)}$ be

$$R_{\text{req}}(n_{s(y)}) = \frac{R_{\text{req}}(G)}{\prod_{x=1}^{y-1} R(n_{s(x)}) \times \prod_{z=y+1}^{|N|} R_{\max}(n_{s(z)})}. \tag{2.15}$$

The reliability requirement of the parallel application is transferred to each task, and the actual reliability must have the following constraint:

$$R(n_{s(y)}) \geqslant R_{\text{req}}(n_{s(y)}). \tag{2.16}$$

However, a main limitation of the above transfer is that the reliability preassignment with maximum reliability value for unassigned tasks is severely pessimistic because of the unfair reliability assignment of tasks, which leads to limited minimization of response times. Given the pessimistic reliability preassignment for unassigned tasks, this section proposes an optimistic reliability preassignment approach.

We first define the reliability rate of the parallel application.

Definition 2.1. *(**Reliability Rate**). The reliability rate of the parallel application is the rate of the reliability requirement of the parallel application to the maximum reliability of the parallel application:*

$$RR(G) = \frac{\sqrt[|N|]{R_{\mathrm{req}}(G)}}{\sqrt[|N|]{R_{\mathrm{max}}(G)}}. \tag{2.17}$$

We obtain $RR(G) \leqslant 1$ because of $R_{\mathrm{req}}(G) \leqslant R_{\mathrm{max}}(G)$ according to Eq. (2.5) in the previous reliability requirement assessment.

Different from MRCRG, in which each unassigned task in $\{n_{s(y+1)}, n_{s(y+2)}, ..., n_{s|N|}\}$ is preassigned to the ECU with the maximum reliability, the rate-based reliability preassignment calculated by Eq. (2.18) is preassigned for these unassigned tasks in this chapter:

$$R_{\mathrm{rrp}}(n_{s(z)}) = R_{\mathrm{max}}(n_{s(z)}) \times RR(G). \tag{2.18}$$

Obviously, we have

$$R_{\mathrm{rrp}}(n_{s(z)}) \leqslant R_{\mathrm{max}}(n_{s(z)}), \tag{2.19}$$

because $RR(G) \leqslant 1$ [see Eq. (2.17)]. Correspondingly, the reliability requirement of task $n_{s(y)}$ is then changed to

$$R_{\mathrm{req}}(n_{s(y)}) = \frac{R_{\mathrm{req}}(G)}{\prod_{x=1}^{y-1} R(n_{s(x)}) \times \prod_{z=y+1}^{|N|} R_{\mathrm{rrp}}(n_{s(z)})}. \tag{2.20}$$

The only difference between Eqs. (2.15) and (2.20) is that $R_{\mathrm{max}}(n_{s(z)})$ of the former is replaced with $R_{\mathrm{rrp}}(n_{s(z)})$ of the latter. In this way, the reliability requirement of the parallel application can still be transferred to each task. When the rate-based reliability preassignment $R_{\mathrm{rrp}}(n_i)$ Eq. (2.15) is adopted, each task n_i in the parallel application can always find an available ECU to meet the reliability requirement $R_{\mathrm{req}}(G)$ of the parallel application. We prove the correctness as follows.

First, we set the product of the rate-based reliability preassignments of all tasks be

$$R_{\mathrm{rrp}}(G) = \prod_{n_i \in N} R_{\mathrm{rrp}}(n_i). \tag{2.21}$$

The preassignment method is feasible if we can prove that the product of rate-based reliability preassignment for all tasks is greater than or equal to the given reliability requirement of the parallel application, and we have proved it.

Second, we substitute Eq. (2.18) into Eq. (2.21) and obtain

$$
\begin{aligned}
R_{\text{rrp}}(G) &= \prod_{n_i \in N} R_{\text{rrp}}(n_i) = \prod_{n_i \in N} (R_{\max}(n_i) \times RR(G)) \\
&= \prod_{n_i \in N} R_{\max}(n_i) \times \prod_{n_i \in N} RR(G).
\end{aligned}
\tag{2.22}
$$

Third, we substitute Eq. (2.17) into Eq. (2.22) and yield

$$
\begin{aligned}
R_{\text{rrp}}(G) &= \prod_{n_i \in N} R_{\max}(n_i) \times \prod_{n_i \in N} \frac{\sqrt[|N|]{R_{\text{req}}(G)}}{\sqrt[|N|]{R_{\max}(G)}} \\
&= R_{\max}(G) \times \frac{R_{\text{req}}(G)}{R_{\max}(G)} = R_{\text{req}}(G).
\end{aligned}
\tag{2.23}
$$

Finally, given that $R_{\text{rrp}}(G)$ is equal to $R_{\text{req}}(G)$ under the rate-based reliability preassignment for tasks in Eq. (2.23), we can find assigned ECUs to meet $R_{\text{req}}(G)$. The reliability requirement of each task is valid. That is, as long as the actual reliability value of $n_{s(y)}$ is set to satisfy the following constraints:

$$
R(n_{s(y)}) \geqslant R_{\text{req}}(n_{s(y)}).
\tag{2.24}
$$

Therefore, when assigning task $n_{s(y)}$, the reliability requirement $R_{\text{req}}(n_{s(y)})$ of $n_{s(y)}$ can be directly considered and the reliability requirement of the parallel application G is not a concern. In this manner, a low time complexity heuristic algorithm can be achieved.

2.4.2 Response Time Minimization

Similar to the HEFT algorithm [58], we also assign n_i to the ECU with the minimum EFT by using the insertion-based scheduling strategy, which is a local optimal heuristic strategy. A local optimum is the smallest EFT that is optimal for the current task when each task is assigned; however, the smallest EFT is not globally optimal for the parallel application. In other words, the local optimum is related to the current task, while the global optimum is related to the parallel application. Given that the reliability requirement of each task is assured in above section, we simply minimize the AFT of each task by traversing all available ECUs under its reliability requirement. Thus, the assigned ECU $u_{\text{pr}(i)}$ for n_i is calculated by

$$
EFT(n_i, u_{\text{pr}(i)}) = \min_{u_k \in U, R(n_i, u_k) \geqslant R_{\text{req}}(n_i)} EFT(n_i, u_k).
\tag{2.25}
$$

In this way, a low complexity heuristic algorithm identical to the HEFT algorithm is implemented. After that, we assign n_i to $u_{\text{pr}(i)}$ based on Eq. (2.25), and the actual AFT and AST of n_i are calculated as follows:

$$
AFT(n_i) = EFT(n_i, u_{\text{pr}(i)}),
\tag{2.26}
$$
$$
AST(n_i) = AFT(n_i) - w_{i,pr(i)}.
\tag{2.27}
$$

Algorithm 1 FFSV1 Algorithm.

Input: $U = u_1, u_2, ..., u_{|U|}$, G, $R_{\text{req}}(G)$, $RT_{\text{req}}(G)$
Output: $RT(G)$, $R(G)$, verification result

1: Order the tasks in a task_list by descending order of $rank_u$ values;
2: Calculate $RR(G)$ using Eq. (2.20);
3: **while** (task_list is not null) **do**
4:     $n_i \leftarrow n_{s(y)} \leftarrow task\_list.out()$;
5:     Calculate $R_{\text{req}}(n_i)$ using Eq. (2.23);
6:     **for** each ECU $u_k \in U$ **do**
7:         Calculate $R(n_i, u_k)$ using Eq. (2.1);
8:         **if** $\mathrm{R}(n_i, u_k) < R_{\text{req}}(n_i)$ **then**
9:             continue;
10:         **end if**
11:         Calculate $EFT(n_i, u_k)$ using Eq. (2.8);
12:     **end for**
13:     Select $u_{\text{pr}(i)}$ with the minimum EFT;
14:     $AFT(n_i) \leftarrow EFT(n_i, u_{\text{pr}(i)})$;
15: **end while**
16: Calculate $R(G)$ using Eq. (2.2);
17: $RT(G) \leftarrow AFT(n_{\text{exit}})$
18: **if** $RT(G) < RT_{\text{req}}(G)$ **then**
19:     **return** true;
20: **else**
21:     **return** false;
22: **end if**

We proposed the first functional safety verification (FFSV1) algorithm, as shown in Algorithm 1. The main idea of FFSV1 is to transfer the reliability requirement of a parallel application to each task using rate-based reliability preallocation, where each task simply selects the ECU with the minimum EFT given the reliability requirement. The details are as follows:

(1) FFSV1 ranks all tasks in a task list by descending order of $rank_u$ values in line 1. FFSV1 calculates the reliability rate using Eq. (2.17) in line 2.

(2) FFSV1 calculates the reliability requirement of the current task n_i using Eq. (2.20) in line 5. Specifically, FFSV1 skips the ECUs that do not meet the reliability requirement of n_i in lines 8−10.

(3) FFSV1 selects the ECU with the minimum EFT for the current task n_i that meets the condition of $R(n_i, u_k) \geqslant R_{\text{req}}(n_i)$ in lines 6−13.

(4) FFSV1 calculates the final $R(G)$ and $RT(G)$ of the parallel application, respectively in lines 16 and 17. FFSV1 judges the verification result by comparing the obtained response time with the given real-time requirement in lines 18–22.

Table 2.5: Upward rank values for all tasks of the motivational parallel application.

n_i	$R_{\text{req}}(n_i)$	$EFT(n_i, u_1)$	$EFT(n_i, u_2)$	$EFT(n_i, u_3)$	$R(n_i)$
n_1	0.995329	-	**16**	-	0.996805
n_2	0.994451	39	**29**	-	0.997403
n_3	0.993972	-	**37**	-	0.998401
n_4	0.990318	**47**	56	-	0.993521
n_5	0.992716	59	**50**	-	0.997403
n_6	0.991933	60	66	**39**	0.991933
n_7	0.994838	-	**75**	-	0.997603
n_8	0.992769	**59**	90	-	0.996506
n_9	0.992588	**69**	86	-	0.997503
n_{10}	0.992209	-	**87**	-	0.998601
		$RT(G) = 87, R(G) = 0.966185 > R_{\text{req}}(G) = 0.96$			

The FFSV1 algorithm's time complexity is analyzed as follows: (1) scheduling all the tasks need traverse all tasks, which can be done within $O(|N|)$ time; (2) calculating EFT value of the current task must traverse its immediate precedence tasks, and all ECUs, which can be done in $O(|N| \times |U|)$ time. Hence, the FFSV1 algorithm's time complexity is $O(|N|^2 \times |U|)$, and it is equal to the time complexity of the HEFT algorithm.

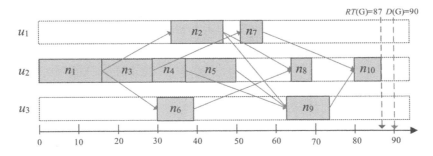

Figure 2.5: Task mapping generated by FFSV1 of the motivational parallel application.

2.4.3 Example of the FFSV1

Table 2.5 shows the task assignment generated by FFSV1 of the motivational parallel application, where each row represents one task assignment. FFSV1 selects the ECU with the minimum EFT and identifies the value in bold. If assigning a task to an ECU does not meet the reliability requirement of the task, the value is indicated by "-". Actually, the reliability value of each task is larger than its reliability requirement. The final reliability is $R(G) = 0.9662$, which can meet the reliability requirement of 0.96. Fig. 2.5 also shows the task mapping generated by FFSV1 of the motivational parallel application satisfies the real-time requirement of 90, where the response time is 87.

The arrows in Fig. 2.5 represent generated communications between tasks. Therefore, the functional safety requirements for the motivational parallel applications can be satisfied by using FFSV1 with a true verification result.

2.5 RELIABILITY MAXIMIZATION UNDER REAL-TIME REQUIREMENT

This section proposes the second functional safety verification approach by maximizing reliability while satisfying response time requirement.

2.5.1 Satisfying Real-Time Requirement

In order to implement fast functional safety verification, we aim to transfer the real-time requirement of the motivational parallel application to each task. The goal of meeting the real-time requirement of the parallel application G is to ensure that all the tasks are finished in $RT_{\text{req}}(G)$ without violating the precedence constraints among tasks. Obviously, the previous real-time requirement assessment using the HEFT algorithm in Fig. 2.4 satisfies the real-time requirement, but the reliability value is excessively low. Hence, we can further enhance the reliability after invoking the HEFT algorithm.

The tasks be can migrated to other ECUs to achieve reliability verification without violating the real-time requirement of the parallel application and the precedence constraints among tasks. Tasks can be migrated to other ECU's slack as long as this migration can be done with higher reliability. In contrast to HEFT and FFSV1 algorithms, the tasks are ordered according to the ascending order of $rank_u$ values to achieve the RT_{req} expansion of tasks. In order to achieve task migration under the precedence constraints among tasks, we first modify the calculations of EST and $RT_{\text{req}}(n_i)$ as follows:

(1) Eq. (2.7) defines the EST and considers the earliest available time $avail[k]$ that the corresponding ECU u_k is in an idle state and ready to execute a task. Since we consider migrating tasks in this section, task n_i should not be limited by the earliest available time. Hence, we update the EST calculation as follows:

$$\begin{cases} EST(n_{\text{entry}}, u_k) = 0, \\ EST(n_i, u_k) = \max_{n_x \in pred(n_i)} \{AST(n_x) + c'_{x,i}\}. \end{cases} \quad (2.28)$$

(2) A greater reliability value is generated only when the task is migrated to another ECU; thus, each task must have individual $RT_{\text{req}}(n_i, u_k)$ on different ECUs, as follows:

$$\begin{cases} RT_{\text{req}}(n_{\text{exit}}, u_k) = RT_{\text{req}}(G), \\ RT_{\text{req}}(n_i, u_k) = \min_{n_j \in succ(n_i)} \{AST(n_j) - c'_{i,j}\}. \end{cases} \quad (2.29)$$

For instance, the ESTs and RT_{req} of n_{10} on all slacks can be obtained as

$$\begin{cases} EST(n_{10}, u_1) = 81, \\ EST(n_{10}, u_2) = 73, \\ EST(n_{10}, u_3) = 81; \end{cases} \qquad \begin{cases} RT_{\text{req}}(n_{10}, u_1) = 90, \\ RT_{\text{req}}(n_{10}, u_2) = 90, \\ RT_{\text{req}}(n_{10}, u_3) = 90. \end{cases} \qquad (2.30)$$

Although EST and RT_{req} are extended to the ECU of each task, task migration must be further limited because each ECU is not always available at all times after using the HEFT algorithm. ECUs remain some slacks because of real-time requirement assessment, and task assignment or task migration is achieved by inserting tasks into ECU. We define the slack set for ECU u_k as follows:

$$S_{i,k} = \{S_{i,k,1}, S_{i,k,2}, S_{i,k,|S_k|}\}, \qquad (2.31)$$

where $S_{i,k,1}$ denotes the first slack on u_k for n_i. The y-th slack $S_{i,k,y}$ is defined as follows:

$$S_{i,k,y} = [t_{\text{s}}(S_{i,k,y}), t_{\text{e}}(S_{i,k,y})], \qquad (2.32)$$

where $t_{\text{s}}(S_{i,k,y})$ represents the start time (ST), and $t_{\text{e}}(S_{i,k,y})$ represents the end time (ET), respectively. As the parallel application has a given real-time requirement, the ET of the last slack must be $t_{\text{e}}(S_{i,k,|S_k|}) = RT_{\text{req}}(G)$. For example, when assigning the task n_{10} in Fig. 2.4, the slacks on u_1, u_2, and u_3 are

$$\begin{cases} S_{10,1} = \{[0, 27], [40, 57], [62, 90]\}, \\ S_{10,2} = \{[0, 18], [42, 56], [68, 90]\}, \\ S_{10,3} = \{[49, 90]\}. \end{cases} \qquad (2.33)$$

Note that n_{10} must be removed from u_2, such that optimal slacks can be selected for it.

Whenever a task needs to meet its real-time requirement and precedence constraints, task n_i should be assigned to the slacks that meet the new EST and RT_{req} constraints:

$$EST(n_i, u_k) = \max\{EST(n_i, u_k), t_{\text{s}}(S_{i,k,t})\}, \qquad (2.34)$$

and

$$RT_{\text{req}}(n_i, u_k) = \min\{RT_{\text{req}}(n_i, u_k), t_{\text{e}}(S_{i,k,t})\}. \qquad (2.35)$$

For instance, the new EST and RT_{req} values of n_{10} on all ECUs are updated to

$$\begin{cases} EST(n_{10}, u_1) = 81, \\ EST(n_{10}, u_2) = 73, \\ EST(n_{10}, u_3) = 81; \end{cases} \qquad \begin{cases} RT_{\text{req}}(n_{10}, u_1) = 90, \\ RT_{\text{req}}(n_{10}, u_2) = 90, \\ RT_{\text{req}}(n_{10}, u_3) = 90. \end{cases} \qquad (2.36)$$

2.5.2 Reliability Maximization

This subsection iteratively allocates each task to the ECU that has the maximum reliability without violating real-time requirement, according to the FFSV algorithm. Since task migration or task assignment is actually task insertion, we must determine whether task insertion is feasible on each ECU prior to task migration. This is because a small slack may not be able to accommodate the insertion of tasks with long WCETs. We make the following constraints before task allocation.

(1) The maximum execution time (MET) for n_i on u_k must be derived as follows:

$$MET(n_i, u_k) = RT_{\text{req}}(n_i, u_k) - EST(n_i, u_k). \tag{2.37}$$

For instance, when assigning task n_{10} according to Fig. 2.4, the METs for n_{10} should be

$$\begin{cases} MET(n_{10}, u_1) = 9 < 21 = w_{10,1}, \\ MET(n_{10}, u_2) = 17 > 7 = w_{10,2}, \\ MET(n_{10}, u_3) = 9 < 16 = w_{10,3}. \end{cases} \tag{2.38}$$

(2) The constraint given as follows must be satisfied, else n_i cannot be inserted into the slack:

$$MET(n_i, u_k) \geqslant w_{i,k}. \tag{2.39}$$

For instance, n_{10} cannot be inserted into u_1 because $MET(n_{10}, u_1) = 9$ is less than $w_{10,1} = 21$, which is the WCET of n_{10} on u_1, shown in Eq. (2.39).

(3) The strategy of maximizing reliability includes the following steps: the real-time requirement of each task has been met in the previous subsection, we can only maximize the reliability value of each task by traversing all available ECUs under the task, which can be inserted into the ECU slack. Thus, the assigned ECU $u_{\text{pr}(i)}$ is determined by

$$R(n_i, u_{\text{pr}(i)}) = \max_{u_k \in U, MET(n_i, u_k) \geqslant w_{i,k}} \{R(n_i, n_k)\}. \tag{2.40}$$

If the reliability values of n_i on other ECUs are less than $R_{\text{heft}}(n_i)$, then n_i is still assigned to the original ECU obtained by the HEFT algorithm. Hence, $R(n_i, u_{\text{pr}(i)})$ must meet the following constraint:

$$R(n_i, u_{\text{pr}(i)}) \geqslant R_{\text{heft}}(n_i), \tag{2.41}$$

where $R_{\text{heft}}(n_i)$ represents the reliability of n_i obtained by HEFT algorithm.

(4) Then, we assign n_i to $u_{\text{pr}(i)}$ based on Eq. (2.40), and the actual AFT and AST of n_i are updated as follows:

$$AFT(n_i) = RT_{\text{req}}(n_i, u_{\text{pr}(i)}), \tag{2.42}$$

and

$$AST(n_i) = RT_{\text{req}}(n_i, u_{\text{pr}(i)}) - w_{i,\text{pr}(i)}. \tag{2.43}$$

Algorithm 2 FFSV2 Algorithm.

Input: $U = u_1, u_2, ..., u_{|U|}$, $G, R_{\text{req}}(G), RT_{\text{req}}(G)$

Output: $RT(G), R(G)$, verification result

1: Invoke the HEFT to obtain the initial allocations of tasks in the parallel application G;
2: Order the tasks in a task_list by ascending order of $rank_u$ values;
3: **while** (task_list) is not null **do**
4:     $n_i \leftarrow task\_list.out()$;
5:     **for** each ECU $u_k \in U$ **do**
6:         Calculate $RT_{\text{req}}(n_i, u_k)$ using Eq. (2.29);
7:         Calculate $MET(n_i, u_k)$ using Eq. (2.39);
8:         **if** $\text{MET}(n_i, u_k) < w_{i,k}$ **then**
9:             continue;
10:         **end if**
11:         Calculate $R(n_i, u_k)$ using Eq. (2.1);
12:     **end for**
13:     Select the $u_{\text{pr}(i))}$ with the maximum reliability value;
14:     $AFT(n_i) \leftarrow RT_{\text{req}}(n_i, u_{\text{pr}(i)})$;
15: **end while**
16: Calculate $R(G)$ using Eq. (2.2);
17: $RT(G) \leftarrow AFT(n_{\text{exit}})$;
18: **if** $R(G) < R_{\text{req}}(G)$ **then**
19:     **return** true;
20: **else**
21:     **return** false;
22: **end if**

We propose the FFSV2 algorithm based on the above analysis, as shown in Algorithm 2. We explain the details of Algorithm 2 as follows.

(1) Obtaining the initial assignments of tasks in the parallel application G by invoking the HEFT algorithm in line 1. FFSV2 orders all tasks in the task_list in ascending order of $rank_u$ values in line 2. In the following, FFSV2 traverseS all tasks.

(2) Calculating the real-time requirement of the current task n_i on all ECUs in line 6. The ECU skipped that n_i cannot be inserted into in lines 8−10.

(3) Selecting the ECU with the maximum reliability value for the current task n_i that satisfies the condition of $MET(n_i, u_k) \geqslant w_{i,k}$ in lines 5−13.

(4) Calculating the final $R(G)$ and $RT(G)$ of the parallel application, respectively in lines 16 and 17. FFSV2 judges the verification result by comparing the obtained reliability value with the given reliability requirement in lines 18−22.

The FFSV2 algorithm's time complexity is also $O(|N|^2 \times |U|)$, which is still equal to the time complexity of the HEFT algorithm. FFSV2 also achieves low-time complexity FFSV.

2.5.3 Example of the FFSV2

Figs. 2.6 and 2.7 show the task mapping generated by FFSV2 of the motivational parallel application. The task n_{10} extends its AST and AFT on u_2 as denoted with shadows in Fig. 2.6. When reassigning n_8, it is migrated from u_2 to u_1 where maximum reliability value of 0.997802 can be obtained without violating the precedence constraints among tasks.

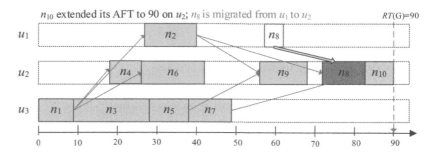

Figure 2.6: Task mapping generated by FFSV2 of n_{10} and n_8.

Finally, the response time value and reliability value of the motivational parallel application are 90 and 0.964737, respectively; and both of the values can meet the corresponding requirements, as shown in Fig. 2.7. Hence, using FFSV2 can also meet the functional safety requirements of the motivational parallel application, and the verification result is true.

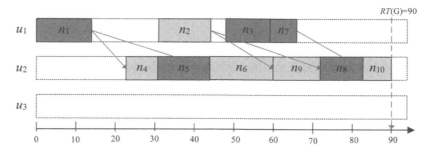

Figure 2.7: Task mapping generated by FFSV2 of the motivational parallel application.

2.5.4 Union Verification

We combine FFSV1 and FFSV2 into UFFSV. The workflow for union verification is shown in Fig. 2.8. The verification returns true whenever either validation method

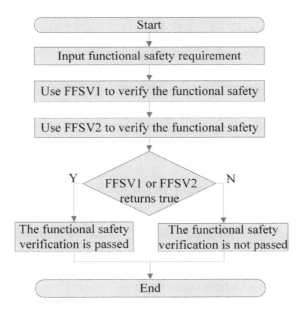

Figure 2.8: Workflow of the UFFSV.

returns true. In any case, both FFSV1 and FFSV2 should be called for the following reasons:

(1) FFSV1 and FFSV2 have different but complementary emphases: FFSV1 discusses minimizing response time with the reliability requirement, while FFSV2 discusses maximizing reliability value with the real-time requirement. Combining these two approaches can achieve the maximum acceptance ratio.

(2) The time complexity of UFFSV is low and does not affect the operational efficiency, because both FFSV1 and FFSV2 are fast verification methods with low time complexity.

(3) Based on their respective verification results, FFSV1 and FFSV2 can guide cost optimization solutions in later design phases.

2.6 EXPERIMENTS FOR FUNCTIONAL SAFETY VERIFICATION ALGORITHMS FFSV2, FFSV2, AND UFFSV

2.6.1 Real-Life Parallel Application

We use the real-life parallel application as shown in Fig. 2.9 from Reference [19, 61]. This real-life parallel application consists of six functional blocks: 1) engine controller with seven tasks ($n_1 - n_7$); 2) automatic gear box with four tasks ($n_8 - n_{11}$); 3) anti-locking brake system with six tasks ($n_{12} - n_{17}$); 4) wheel angle sensor with two tasks ($n_{18} - n_{19}$); 5) suspension controller with five tasks ($n_{20} - n_{24}$); and 6) body work with seven tasks ($n_{25} - n_{31}$). The number of ECUs in this embedded system is 16, and the

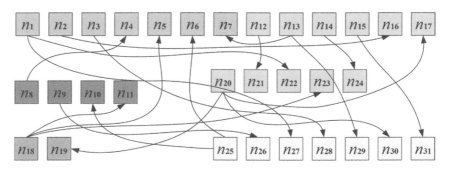

Figure 2.9: Real automotive parallel application [19, 61].

failure rate per task is in the range of $0.000001/\mu s$ $-0.000016/\mu s$. The WCETs of the tasks and the WCRTs of the messages are in the range of 100 μs $-400\,\mu s$. Considering that two verification algorithms are proposed in this chapter, we first compare these two algorithms with their existing counterparts.

Experiment 1: The experiment compares the response time values of real parallel applications with different reliability requirements. To the best of our knowledge, MRCRG is a state-of-the-art approach to minimizing the cost of resource consumption while meeting the reliability requirement of the parallel application on embedded systems [61]. We compare this approach with FFSV1 and name it as MRTP (minimizing response time with pessimism) in this section. Considering the $R_{\max}(G) = 0.986271$, the reliability requirement is changed from 0.9 to 0.98 with a 0.01 increment; $0.9-0.98$ fall within the range of exposure E3 in ISO 26262. Considering that the actual reliability value is always greater than or equal to the reliability requirement corresponding to the use of MRTP and FFSV1 , we do not list the reliability values in this experiment. Fig 2.10 shows the response time values for different reliability requirements, the x-axis and y-axis represent the reliability requirement and the actual response time, respectively.

(1) While response time values do not increase proportionally with reliability requirement, increasing reliability will increase overall response time, especially when reliability requirement are high. The curve shows that response time minimization and reliability maximization are often in conflict with each other.

(2) According to the analysis of the relationship between response time and reliability, the curve generated by FFSV1 in Fig. 2.10 should be Pareto optimal. However, this is not the case. We can see that the reliability requirement from 0.9 to 0.92 yields a higher response time value than that of 0.93. Since shortening the development cycle is extremely critical for development cost control in the automotive industry, the FFSV1 algorithm sacrifices accuracy to ensure the development schedule in order to save time, which affects the accuracy of the results. Hence, FFSV1 cannot ensure the Pareto optimal curve in Fig. 2.1 can be obtained.

(3) In all cases, the response time values generated by FFSV1 are shorter (or equal)

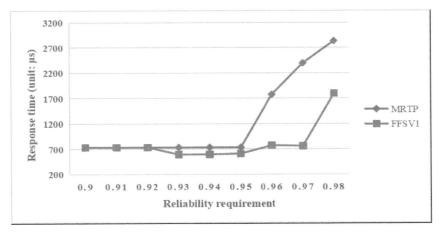

Figure 2.10: Response time values of the real-life application under different reliability requirements.

than those of MRTP. When the reliability requirement is greater than 0.95, FFSV1 outperforms MRTP in minimizing response time values. When the reliability requirement is 0.97, FFSV1 can reduce the response time by as much as 68.06% compared to MRTP because of the relatively optimistic reliability preassignment of FFSV1 for unfair reliability usage among tasks; therefore, FFSV1 can generate a shorter response time value than MRTP. Hence, FFSV1 is identified as a good choice for the first type of functional safety verification (i.e., minimizing response time while meeting reliability requirement).

Experiment 2: This experiment compares reliability values for the real-life parallel application with different real-time requirements. To the best of our knowledge, there are no studies similar to the FFSV2 approach. We know that the HEFT algorithm is a well-studied parallel application assessment algorithm that consistently meets the real-time requirements. Hence, we use HEFT to compare with FFSV2. The lower-bound of the parallel application is $LB(G) = 736$ μs. Therefore, the real-time requirement is changed from 736 to 1536 μs with 100 μs increments, as shown on the x-axis in Fig. 2.11. Considering that the actual response time values when using the HEFT algorithm and FFSV2 are always less than or equal to the corresponding real-time requirements, we do not list the response time values in this experiment. The reliability values for different real-time requirements are shown in Fig 2.11, where the x-axis and y-axis represent the real-time requirements and the actual reliability value, respectively.

(1) As expected, the reliability values generated by HEFT are fixed and do not change as the real-time requirement increases. Because it is only designed to obtain a lower bound and does not involve reliability enhancement.

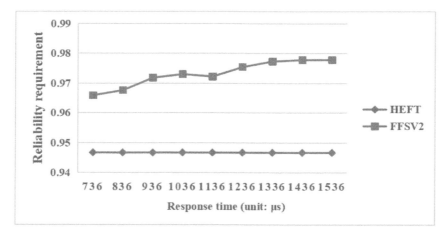

Figure 2.11: Reliability values of the real-life parallel application under different real-time requirements.

(2) In all cases, FFSV2 consistently generates higher reliability values than the HEFT algorithm. The difference in reliability between the HEFT algorithm and FFSV2 algorithm reaches 0.03126 when the real-time requirement is 1536 μs. The increase in the FFSV2 curve indicates that higher real-time requirement can create larger slacks, which leads to greater reliability values. The increased reliability values of the FFSV2 algorithm confirm that FFSV2 is a good choice for functional safety verification.

2.6.2 Synthetic Parallel Application

In view of the increasing complexity of automotive embedded systems, future automotive parallel applications may include at least 50 tasks and up to 100 tasks. To further verify the effectiveness, we observe the results using other synthetic parallel applications with the same actual parameter values as the actual parallel application. Randomly parallel applications can be generated by a task graph generator. The number and failure rate of ECUs, the WCETs of tasks and the WCRTs of messages are the same as for real-life parallel applications.

Experiment 3: This experiment shows the acceptance rates of a synthetic application using six verification approaches for various tasks. We set the parallel application parameters as follows: the average WCET is 200 μs, the communication-to-computation ratio is 1, the shape parameter is 1, and the heterogeneity factor is 0.5. Heterogeneity factor values are in the range 0−1 in the task graph builder, where 0.1 and 1 are the lowest and highest heterogeneity factors, respectively. The number of tasks is changed from 50 to 100, in increments of 10. The reliability requirement is changed from 0.9 to the maximum reliability value in increments of 0.01. The real-time requirement is changed from lower-bound values with 100 μs with the same number iteration of reliability requirement. Table 2.6 illustrates the acceptance rates of synthetic parallel applications for different number of tasks using the six methods.

Table 2.6: Acceptance rates of synthetic applications for different numbers of tasks

Approach	$|N| = 50$	$|N| = 60$	$|N| = 70$	$|N| = 80$	$|N| = 90$	$|N| = 100$
MRTP	75%	73%	57%	20%	31%	28%
HEFT	75%	50%	57%	14%	17%	0%
BSH	86%	72%	63%	54%	44%	40%
FFSV1	89%	63%	63%	49%	42%	42%
FFSV2	88%	73%	71%	53%	50%	47%
UFFSV	**92%**	**80%**	**76%**	**57%**	**56%**	**50%**

(1) The acceptance rates of MRTP and HEFT are lower than those of FFSV1 and FFSV2, respectively, for different task numbers. The advantages of FFSV1 and FFSV2 over MRTP and HEFT are evident. One special case is that when the number of tasks is 100, the acceptance rate using the HEFT is 0, because the HEFT only obtains a maximum reliability of 0.893546, which is less than 0.9.

(2) In most cases, the acceptance rate of FFSV2 is slightly higher than that of FFSV1, but the results are not absolute. We cannot determine which is better, FFSV1 or FFSV2, because they are dual problems that use completely different ideas.

(3) The UFFSV has higher acceptance rates than FFSV1 and FFSV2 alone. Hence, the UFFSV is definitely better than individual independent verification. UFFSV can improve acceptance rates by up to 57% (when $N = 80$) and 50% (when $N = 100$), over the MRTP and HEFT, respectively. In conclusion, the UFFSV shows good results compared to other similar methods in both real-life and synthetic parallel applications.

2.7 CONCLUDING REMARKS

In this chapter, we propose two fast methods for functional safety verification. One approach, called FFSV1, finds solutions with minimum response time under reliability requirements, and the second approach, called FFSV2, finds solutions with maximum reliability under real-time requirements. We combine FFSV1 and FFSV2 into a union FFSV (UFFSV). The verification returns true whenever either verification method returns true.

UFFSV is a fast heuristic that shortens the parallel application development life-cycle. Experimentally, UFSFV can improve the acceptance rate over existing comparable methods. If the verification fails, the analyst needs to reanalyze the factors affecting the verification results. If the verification passes, the designer can propose cost optimization solutions based on the corresponding verification scenarios. Therefore, in our future work, if the verification fails, we will improve the system analysis to meet the intended verification requirement, and if the verification passes, we will perform cost design optimization based on the UFFSV results.

CHAPTER 3

Functional Safety Enhancement

S AFETY enhancement is necessary for risk control in embedded systems. This chapter devotes to functional safety enhancement for a parallel application of embedded systems under real-time requirement. We present a stable stopping-based functional safety enhancement (SSFSE) method for a parallel application of embedded systems based on the static recovery mechanism provided in ISO 26262. The SSFSE method combines known backward recovery (BFSE), presented forward recovery (FFSE), and presented forward and backward recovery (RFFSE and RBFSE) through primary-backup repetition. The SSFSE method is a convergence algorithm, which means that the algorithm can stop when the reliability value reaches a steady state. At the end of this chapter, we design different experiments. The results of experiments reveal that the exposure level defined in ISO 26262 decreases from E3 to E1 after using SSFSE method, and such improvement leads to a higher level of safety assurance.

3.1 INTRODUCTION

Safety is a concern throughout the lifecycle of embedded system development, and the maximum possible safety values should be known in the early design phase of the embedded system to help control the risks during actual development progress. Safety enhancement (i.e., safety mechanisms for error handling) is an extremely effective method for risk control.

Risk means the probability of harm occurring and the severity of the harm [28]. Therefore, safety can be enhanced by reducing the possibility or severity of harm or both to reduce risk. In ISO 26262, severity is a measure of the degree of harm caused to an individual in a particular situation [28] and is fixed for a specific automotive parallel application as it is determined by the nature of the harm. Therefore, the only way to reduce risk is to reduce the probability of harm occurring. In ISO 26262, the probability of harm occurring is expressed in terms of exposure [28]. Reliability is commonly related to random hardware failure and is used to express the probability of no harm occurring (i.e., reliability = 1 - exposure) in the automotive functional safety field.

A real-time parallel application must ensure correct response within its deadline [47]. In automotive embedded systems, many safety-critical parallel applications have tight time constraints such as anti-lock brakes [2]. If the anti-lock brakes cannot meet its deadline, vehicle collisions may occur due to the delay of the brake action output, resulting in injury or harm to people (including drivers, passengers, and pedestrians) or damage to vehicles and roadway. The response time of the real-time parallel application exceeding its deadline is one of the system faults that can lead to malfunctioning behavior. That is, the parallel application of an automotive embedded system must be achieved correctly within its deadline; otherwise, the probability of injury occurrence is considered to be 100%. Hence, the essence of functional safety enhancement is to enhance the reliability of the parallel application of automotive embedded systems while ensuring its real-time constraints. Reference [60] presented the reliability enhancement technique (RET) to enhance the functional safety of a real-time parallel application of an automotive embedded system. RET is a backward functional safety enhancement (BFSE) algorithm as it attempts to migrate each task to another ECU, yielding the maximum reliability value (i.e., backward recovery) from the exit task to the entry task (details on BFSE are given in Section 3.4.1). However, the use of BFSE alone is not sufficient to enhance functional safety for the following reasons:

(1) BFSE only enhances reliability by backward recovery (from the exit task to the entry task) and does not apply forward recovery (from the entry exit task to the exit task). For parallel applications of automotive embedded systems, it may be effective to apply forward and backward recovery to enhance functional safety.

(2) BFSE is an unduplicated approach and therefore severely limits the strength of safety enhancement. Actually, safety can be achieved by repeated recovery, an effective fault-tolerance measure using redundancy.

This chapter introduces functional safety enhancement techniques for a parallel application of an automotive embedded system by using fault tolerance measure during the design phase. The main contributions of this chapter are summarized as follows:

(1) We devise the forward safety enhancement (FSE) algorithm for parallel applications of automotive embedded systems. In contrast to BFSE , FFSE addresses the recovery process from entry task to exit task. And FFSE tries to forward reassign each task to another ECU that can obtain maximum reliability value while meeting functional safety requirements. Then we propose the forward and backward recovery to further enhance functional safety by combining BFSE and FFSE algorithms.

(2) We propose the repeating BFSE (RBFSE) algorithm and the repeating FFSE (RFFSE) algorithm for better safety enhancement. Unlike BFSE and FFSE which do not repeat the recovery process, RBFSE and RFSE repeat the recovery process by the primary backup. RBFSE (or RFSE) is a novel contribution because it tries to backward (or forward) add a new copy of each task to the available ECUs, which yields the maximum reliability value among all available ECUs without violating the given constraint.

(3) We present the stable stopping-based functional safety enhancement (SSFSE) method by combining the BFSE, FFSE, RBFSE, and RFFSE algorithms. The SSFSE method is a convergent algorithm, which means that the algorithm can stop when the reliability value reaches a steady state. SSFSE method is a novel contribution because it is a new combination of methods.

3.2 RELATED WORK

This chapter aims to enhance functional safety for the parallel application of an embedded system by considering two safety properties, namely, reliability and response time. This section mainly reviews existing research on the reliability and response time of the parallel application.

(1)Bi-objective optimization. Simultaneous minimization of response time and maximization of reliability is a bi-objective optimization problem [9, 32, 53]. Reference [9] proposed the NP-hard complexity results and the optimal mapping algorithm for the bi-objective optimization problem under different variants of multiprocessors. Reference [32] provided a meta-heuristic algorithm to optimize the bi-objective problem for a parallel application while satisfying the user-defined budget. Reference [53] proposed a bi-objective optimization algorithm for a parallel application based on Wind-Driven Optimization (WDO) to implement the trade-off between minimization of response time and maximization of reliability.

(2) Optimizing response time under reliability requirement. Reference [60] proposed a method to achieve the response time minimization of a parallel application while ensuring the reliability requirement without primary-backup repetition. References [75, 76] proposed the RR and MaxRe algorithms to achieve the resource cost minimization of a parallel application while ensuring the reliability requirement by primary-backup repetition. To effectively ensure the reliability requirement, References [75, 76] presented non-repeated and repeated methods.

(3) Optimizing reliability under real-time requirement. Reference [60] first focused on enhancing the functional safety for the parallel application of an automotive embedded system while ensuring its real-time requirement through proposing the BFSE algorithm. We explain the BFSE algorithm in Section 3.1, it is only a backward safety enhancement algorithm, thus severely limiting the strength of the functional safety enhancement. In other words, the BFSE algorithm is not sufficient to enhance algorithm safety. The purpose of this chapter is to present a new functional safety enhancement method by describing backward and forward recovery and primary backup repetition for the parallel application of an embedded system.

3.3   MODELS AND PROBLEM STATEMENT

We adopt the same parallel application model and reliability model in Section 2.3.1. Tables 3.1 and 3.2 list the abbreviations and notations that are used in this chapter.

Table 3.1: Abbreviations in this chapter.

Abbreviation	Definition
WDO	Wind Driven Optimization
OCT	Optimistic Cost Table
ECC	Error Correcting Code
RET	Reliability Enhancement Technique
MRI	Message Receiving Interrupt
ISR	Interrupt Service Routine
BFSE	Backward Functional Safety Enhancement
RBFSE	Repeated Backward Functional Safety Enhancement
FFSE	Forward Functional Safety Enhancement
RFFSE	Repeated Forward Functional Safety Enhancement
SSFSE	Stable Stopping-based Functional Safety Enhancement

Table 3.2: Notations in this chapter.

Notation	Definition
$D(G)$	Deadline of the parallel application G
$LFT(n_i, u_k)$	Latest finish time of task n_i executed in ECU u_k
$avaST(n_i, u_k)$	Available start time of task n_i executed in ECU u_k
$avaET(n_i, u_k)$	Available end time of task n_i executed in ECU u_k

3.3.1   Lower Bound of Application

Definition 3.1. (**Lower bound**). *The lower bound refers the minimum response time of a parallel application without any constraints.*

In this chapter, we also apply the HEFT to calculate the lower bound of the parallel application. The main ideas of HEFT are as follows:

(1) Task priority is determined based on the descending order of $rank_u$, which is calculated by Eq. (2.6) [58]. In the parallel application of Fig. 2.3, the task priority is organized by n_1 ($rank_u(n_1) = 108$), n_3 ($rank_u(n_3) = 80$), n_4 ($rank_u(n_4) = 80$), n_2 ($rank_u(n_2) = 77$), n_5 ($rank_u(n_5) = 69$), n_6 ($rank_u(n_6) = 63.3$), n_9 ($rank_u(n_9) = 44.3$), n_7 ($rank_u(n_7) = 42.7$), n_8 ($rank_u(n_8) = 35.7$), and n_{10} ($rank_u(n_{10}) = 14.7$) by using the HEFT [58].

(2) The task assignment can obtain the lower bound of the parallel application. Let $EFT(n_i, u_k)$ be the EFT of task n_i executed in ECU u_k. The minimum EFT of the exit task should be the lower bound of the motivational parallel application:

$$LB(G) = \min_{u_k \in U} \{EFT(n_{\text{exit}}, u_k)\}. \tag{3.1}$$

Hence, the minimum EFT for each task can be obtained iteratively from the entry task to the exit task. The EFT for each task is calculated by

$$EFT(n_i, u_k) = EST(n_i, u_k) + w_{i,k}, \tag{3.2}$$

where $EST(n_i, u_k)$ denotes the EST of task n_i executed in ECU u_k:

$$\begin{cases} EST(n_{\text{entry}}, u_k) = 0, \\ EST(n_i, u_k) = \max \left(avaST(n_i, u_k), \max_{n_h \in pred(n_i)} \{AFT(n_h) + c_{h,i}^{p,k}\} \right); \end{cases} \tag{3.3}$$

$avaST(n_i, u_k)$ is the available start time of u_k, and n_h is an immediate predecessor task of n_i. $AFT(n_h)$ is the AFT of n_h, and n_i has different $EST(n_i, u_k)$ according to the assignments of predecessor tasks since $c_{h,i}^{p,k}$ is not a fixed value.

$$c_{h,i}^{p,k} = \begin{cases} c_{h,i} & p \neq k. \\ 0 & p = k. \end{cases} \tag{3.4}$$

Table 3.3 shows the lower bound of the motivational parallel application G. First, n_1 is assigned to u_3 because the assignment has the minimum EFT of 9. Then, n_3 is assigned to u_3 because it has the minimum EFT of 28. The EFTs of n_4, n_2, n_5, n_6, n_9, n_7, n_8, and n_{10} are calculated and shown in Table 3.3, and the lower bound of the motivational parallel application G is 80 because the AFT of n_{10}(i.e., the exit task) is 80. Fig. 3.1 shows the lower bound generated by HEFT of the motivational parallel application G in Fig. 2.3. The deadline of the motivational parallel application must be larger than or equal to $LB(G)$. For the motivational parallel application in Fig. 2.3, we let the deadline be $D(G) = 100$.

Table 3.3: Lower bound of the motivational parallel application.

Task	$EFT(n_i, u_1)$	$EFT(n_i, u_2)$	$EFT(n_i, u_3)$
n_1	14	16	**9**
n_3	32	34	**28**
n_4	31	**26**	45
n_2	**40**	46	46
n_5	52	39	**38**
n_6	53	**42**	47
n_9	69	**68**	76
n_7	58	83	**49**
n_8	**62**	79	73
n_{10}	102	**80**	97

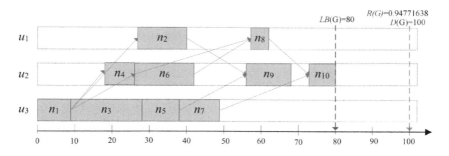

Figure 3.1: Lower bound generated by the HEFT algorithm.

3.3.2 Problem Statement

Given that the essence of functional safety enhancement is to enhance the reliability of the parallel application of the embedded system while ensuring its real-time requirement as explained in Section 3.1, We mainly work on enhancing the functional safety of the motivational parallel application G in this chapter.

$$R(G) = \prod_{n_i \in N} R\left(n_i\right), \tag{3.5}$$

enhancing functional safety through task assignment, under real-time requirement

$$RT(G) \leqslant D(G), \tag{3.6}$$

where $RT(G)$ represents the response time of the motivational parallel application G.

3.4 BACKWARD AND FORWARD SAFETY ENHANCEMENT

3.4.1 Existing BFSE Algorithm

The BFSE algorithm is a backward functional safety enhancement method. Its main idea is to try to reassign the current task n_i to another ECU that generates maximum reliability value of n_i while meeting real-time requirement. The backward recovery means from the exit task to the entry task. The following is a brief explanation of the BFSE idea based on the lower bound generated by the HEFT algorithm (i.e., Fig. 3.1):

(1) The backward sequences of the tasks are arranged in descending order of the AFT values generated by HEFT. For instance, the backward sequence of tasks in Fig. 3.1 is n_{10}, n_9, n_8, n_7, n_6, n_2, n_5, n_3, n_4, and n_1.

(2) We first consider the exit task n_{10}. Moving the end time of n_{10} to 100 (i.e., $D(G) = 100$) in fixed ECU u_2 while meeting real-time requirement, as shown in Fig. 3.2. We assume that n_{10}, n_9, n_8, and n_7 have been reassigned by using BFSE, as shown in Fig. 3.2. Note that n_{10}, n_9, n_8, and n_7 are only re-scheduled on the same ECU without migration. Task migration is not feasible for the

following reasons: 1) if task migration would result in a violation of precedence constraints, such task migration is not allowed. 2) if task migration does not obtain higher reliability values, such migration is not allowed.

(3) Next, task n_6 is considered to be reassigned. n_6 can be migrated from u_2 to u_3 because this migration does not violate the priority constraint and obtains the maximum reliability value for n_6, as shown in Fig. 3.2.

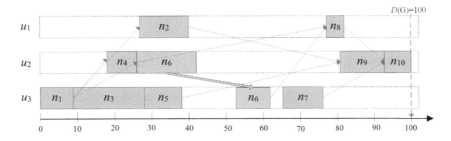

Figure 3.2: Task migration by using the BFSE algorithm (n_{10}, n_9, n_8, and n_7 are not migrated, and n_6 is migrated from u_2 to u_3).

(4) The remaining tasks use the same strategy as the previous tasks $n_6 - n_{10}$. In tasks $n_1 - n_5$, only task n_3 is migrated from u_3 to u_2; n_5, n_4, n_2, n_1 cannot be migrated to other ECU and they can only be moved in fixed ECUs. Finally, the reliability of the parallel application G is enhanced from 0.9477163 to 0.95294318 by using RET to migrate n_6 and n_3 as shown in Figs. 3.1 and 3.3, respectively. Fig. 3.3 shows that precedence constraint and real-time requirement can be satisfied.

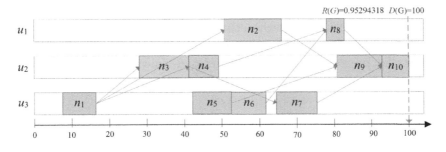

Figure 3.3: Lower bound generated by BFSE of the motivational parallel application.

3.4.2 FFSE Algorithm

The start time of the task n_1 does not start from 0 but from 8 after using BFSE as shown in Fig. 3.3. An intuition is that by performing some possible migrations for n_1 and its successor tasks, the start time of n_1 can be set to 0, thus enhancing functional safety while satisfying precedence constraint and real-time requirement. The above process explores from the entry task to the exit task (i.e., forward recovery).

(1) The forward sequence of tasks is ordered by the ascending order of AST values calculated by the BFSE algorithm. For instance, the forward sequence of tasks in Fig. 3.3 is n_1, n_3, n_4, n_5, n_2, n_6, n_7, n_8, n_9, and n_{10}.

(2) The n_1 is in the fixed ECU u_3 without migration, and its start time and end time are shifted to 0 and 8, respectively. In Section 3.4.1 we have verified that task n_1 cannot be migrated, as shown in Fig. 3.3.

(3) n_1, n_3, n_4, and n_5 have shifted their individual start times and end times to smaller values in fixed ECUs without migrating to other ECUs, as shown in Fig. 3.4.

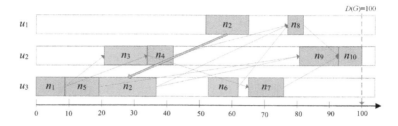

Figure 3.4: n_1, n_3, n_4, and n_5 have moved their individual start times and end times to small values in fixed ECUs without migrating to other ECUs.

(4) The task n_2 is ready to be reassigned. The n_2 can be migrated from u_3 to u_1 because this migration can obtain maximum reliability for n_2 while meeting precedence constraints, as shown in Fig. 3.4. Although the n_2 cannot be migrated by using BFSE, n_2 can be migrated upon further use of FFSE. The migration of n_2 will be explained as follows.

In order to determine whether the migration of the current task n_i can be performed, its EST and LFT in each ECU need to be obtained in advance to ensure that precedence constraints and real-time requirement are satisfied. The EST of n_i in u_k has been shown in Eq. (3.3), whereas the LFT of n_i in u_k is calculated by

$$\begin{cases} LFT(n_{\text{exit}}, u_k) = D(G); \\ LFT(n_i, u_k) = \min\left(avaET(n_i, u_k), \min_{n_j \in succ(n_i)}\{AST(n_j) - c_{i,j}^{k,q}\}\right); \end{cases} \quad (3.7)$$

$avaET(n_i, u_k)$ is the available end time of u_k for n_i. n_i has different $LFT(n_i, u_k)$ values according to the assignments of successor tasks because $c_{i,j}^{k,q}$ is not a fixed value.

$$c_{i,j}^{k,q} = \begin{cases} c_{i,j}, & k \neq q. \\ 0, & k = q. \end{cases} \quad (3.8)$$

Let's take n_2 as an example. The $avaST(n_2, u_k)$ and $avaET(n_2, u_k)$ values are as follows:

$$\begin{cases} avaST(n_2, u_1) = 0, \\ avaST(n_2, u_2) = 42, \\ avaST(n_2, u_3) = 19; \end{cases} \qquad \begin{cases} avaET(n_2, u_1) = 77, \\ avaET(n_2, u_2) = 81, \qquad (3.9) \\ avaET(n_2, u_3) = 53. \end{cases}$$

Then, the $EST(n_2, u_k)$ and $LFT(n_2, u_k)$ values are as follows:

$$\begin{cases} EST(n_2, u_1) = 27, \\ EST(n_2, u_2) = 42, \\ EST(n_2, u_3) = 19; \end{cases} \qquad \begin{cases} LFT(n_2, u_1) = 65, \\ LFT(n_2, u_2) = 58, \qquad (3.10) \\ LFT(n_2, u_3) = 53. \end{cases}$$

After precedence constraint and real-time requirement are determined, the ECU u_{pr} that has the maximum reliability value for n_i can then be selected.

$$R(n_i) = R(n_i, u_{\text{pr}(i)}) = \max_{w_{i,k} \leqslant (LFT(n_i, u_k) - EST(n_i, u_k))} R(n_i, u_k), \qquad (3.11)$$

where $w_{i,k} \leqslant (LFT(n_i, u_k) - EST(n_i, u_k))$ should be satisfied to ensure no violations of precedence constraint. Note that u_{pr} can be the currently assigned ECU u_{cur} to be moved in or the reassigned ECU u_{new} to be migrated to.

Continuing with n_2 as an example. The $R(n_2, u_k)$ values are as follows:

$$\begin{cases} R(n_2, u_1) = 0.99094128, \\ R(n_2, u_2) = \text{NULL}, \qquad\qquad (3.12) \\ R(n_2, u_3) = 0.99282586. \end{cases}$$

$R(n_2, u_2)$ is NULL because $w_{2,2} = 19$, which is larger than 16 ($LFT(n_2, u_2) - EST(n_2, u_2) = 58 - 42 = 16$), such that allocating n_2 to u_2 will violate precedence constraint. Given that $R(n_2, u_3)$ has the maximum reliability values in Eq. (3.12), n_2 is migrated from u_1 to u_3.

The AST and AFT of n_i are correspondingly updated to

$$\dot{A}ST(n_i) = EST(n_i, u_{\text{pr}(i)}), \qquad (3.13)$$

and

$$AFT(n_i) = AST(n_i) + w_{i,\text{pr}(i)}, \qquad (3.14)$$

respectively. For example, the AST and AFT of n_2 are as follows:

$$AST(n_2) = EST(n_2, u_3) = 19, \qquad (3.15)$$

and

$$AFT(n_2) = AST(n_2) + w_{2,3} = 19 + 18 = 37. \qquad (3.16)$$

(5) The following tasks use the same principle as the above tasks. As shown in Fig. 3.5, n_6, n_7, n_8, n_9, and n_{10} are moved in fixed ECUs without migration. Finally, the reliability of the parallel application is enhanced from 0.95294318 (Fig. 3.3) to 0.95475549 (Fig. 3.5) by migrating n_2 using FFSE. Fig. 3.5 shows that precedence constraint and real-time requirement are still not violated.

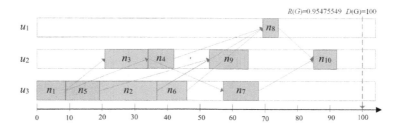

Figure 3.5: Task mapping generated by FFSE of the motivational parallel application.

We propose the FFSE algorithm on the basis of the aforementioned analysis, as shown in Algorithm 3. In general, FFSE attempts to reassign each task to another ECU that can produce the maximum reliability value without violating precedence constraint and real-time requirement.

Algorithm 3 FFSE Algorithm

Input: $U = \{u_1, u_2, ..., u_{|U|}\}$, G, $D(G)$, task mapping generated by the previous algorithm

Output: $R(G)$, $RT(G)$, and task mapping

1: Sort $forward\_seq$ by the ascending order of the AST values generated from the previous algorithm;
2: **while** ($forward\_seq$ is not NULL) **do**
3:     $n_i \leftarrow forward\_seq.out()$;
4:     $u_{\mathrm{cur}(i)}$ indicates the currently assigned ECU.
5:     Move the n_i in ECU $u_{\mathrm{cur}(i)}$ by $AST(n_i)$ and $AFT(n_i)$;
6:     **for** (each ECU $u_k \in U$) **do**
7:         Calculate $EST(n_i, u_k)$ using Eq. (3.3);
8:         Calculate $LFT(n_i, u_k)$ using Eq. (3.7);
9:         **if** $(((LFT(n_i, u_k) - EST(n_i, u_k)) < w_{i,k})$ **then**
10:             **continue**;
11:         **end if**
12:         Calculate $R(n_i, u_k)$ using Eq. (2.1);
13:     **end for**
14:     Select the $u_{\mathrm{new}(i)}$ with the max $R(G)$ under $R(n_i, u_{\mathrm{new}(i)}) > R(n_i, u_{\mathrm{cur}(i)})$;
15:     **if** ($u_{\mathrm{new}(i)}$ is not NULL) **then**
16:         Migrate the n_i to $u_{\mathrm{new}(i)}$ by $AST(n_i)$ and $AFT(n_i)$;
17:     **end if**
18: **end while**
19: Calculate $R(G)$ using Eq. (2.2).

The FFSE algorithm's time complexity is $O(|N|^2 \times |U|)$ and is explained below. (1) traversing all tasks requires $O(|N|)$ time in lines 2−18. (2) calculating $EST(n_i, u_k)$ and $LFT(n_i, u_k)$ requires $O(|N| \times |U|)$ time (lines 6−11). Hence, the time complexity of the FFSE algorithm is the same as those of HEFT and BFSE.

3.5 REPEATED SAFETY ENHANCEMENT

3.5.1 RBFSE Algorithm

Fig. 3.5 shows that the end time of the exit task n_{10} is not ended at 100 but at 92 after using FFSE algorithm. The intuition is that the end time of n_{10} can be set to 100 for some possible task migration via BFSE again. It is necessary to set the end time of n_{10} to 100, but it is not feasible to migrate tasks through BFSE again. The reason is that the task migration reaches its limit through the BFSE and FFSE algorithms in Section3.4. Continuous invocation of BFSE or FFSE recovery cannot generate new migrations. Fortunately, primary backup repetition is an effective fault tolerance measure to achieve functional safety enhancement (i.e., recovery by repetition).

Passive and active replication are the two main backup replication paradigms [75, 76]. Passive replication is designed to reschedule tasks on a backup ECU when a task in the primary ECU fails. Active replication will execute a copy of the task in the ECU at the same time. Each ECU performs only one copy of the same task, so the maximum number of copies is the number of ECUs. Homogeneous redundancy and heterogeneous redundancy are two types of redundancy examples [30]. Homogeneous redundancy implies the replication of homogeneous elements, while heterogeneous redundancy implies the combination of hardware devices and software tasks. This chapter uses active repetition and homogeneous redundancy, as this combination achieves independence between tasks. In addition, as noted in ISO 26262 Version 2, homogeneous redundancy in the design phase focuses on controlling the effects of transient or random failures in the hardware (i.e., the ECUs) by performing similar software on the hardware (e.g., time-redundant implementation of software). The above characteristics are consistent with the parallel application and reliability model in this chapter.

In view of the reliability of n_i in u_k is $R\left(n_i, u_k\right) = e^{-\lambda_k w_{i,k}}$ according to Eq. (2.1), the failure probability of n_i in u_k without repetition is

$$Fail(n_i, u_k) = 1 - R\left(n_i, u_k\right) = 1 - e^{-\lambda_k w_{i,k}}. \tag{3.17}$$

Assume that there are num_i ($num_i \leqslant |U|$) copies for n_i, the failure probability of n_i through active repetition is

$$
\begin{aligned}
Fail(n_i) &= \prod_{\beta=1}^{num_i} Fail(n_i^\beta, u_{\mathrm{pr}(i)(n_i^\beta)}) \\
&= \prod_{\beta=1}^{num_i} \left(1 - R\left(n_i^\beta, u_{\mathrm{pr}(i)(n_i^\beta)}\right)\right),
\end{aligned} \tag{3.18}
$$

where $u_{\mathrm{pr}(i)(n_i^\beta)}$ represents the assigned ECU of copy n_i^β. Hence, the reliability of n_i is

$$R\left(n_i\right) = 1 - Fail(n_i) = 1 - \prod_{\beta=1}^{num_i} \left(1 - R\left(n_i^\beta, u_{\mathrm{pr}(i)(n_i^\beta)}\right)\right). \tag{3.19}$$

(1) Sorting the backward sequence of tasks by the descending order of the AFT values. For instance, the backward sequence of the tasks in Fig. 3.5 is n_{10}, n_8, n_7, n_9, n_6, n_4, n_2, n_3, n_5, and n_1.

(2) First, we consider the exit task n_{10}. n_{10} is just moved its start time to 93 and end time to 100 in fixed ECU u_2 without any repetition. Repetitions on u_1 or u_3 will violate precedence constraint. The details are explained below. The EST and LFT of n_{10} are

$$\begin{cases} EST(n_{10}, u_1) = 85, \\ EST(n_{10}, u_2) = 85, \\ EST(n_{10}, u_3) = 85; \end{cases} \qquad \begin{cases} LFT(n_{10}, u_1) = 100 \\ LFT(n_{10}, u_2) = 100 \\ LFT(n_{10}, u_3) = 100. \end{cases} \qquad (3.20)$$

The LFT values of n_{10} on u_1 and u_3 are

$$LFT(n_{10}, u_1) - EST(n_{10}, u_1) = 100 - 85 = 15 < w_{10,1} = 21, \qquad (3.21)$$

and

$$LFT(n_{10}, u_3) - EST(n_{10}, u_3) = 100 - 85 = 15 < w_{10,3} = 16; \qquad (3.22)$$

hence, neither u_1 nor u_3 can be added with copies.

(3) The task n_8 is second considered to be repeated. A copy can be added to u_2 for n_8 because this operation does not violate precedence constraint, as shown in Fig. 3.6. We explain why a copy can be added to u_2 for n_8 in the following. The EST and LFT values of n_8 are

$$\begin{cases} EST(n_8, u_1) = 69, \\ EST(n_{10}, u_2) = 65, \\ EST(n_8, u_3) = 69; \end{cases} \qquad \begin{cases} LFT(n_8, u_1) = 82, \\ LFT(n_8, u_2) = 93, \\ LFT(n_8, u_3) = 82. \end{cases} \qquad (3.23)$$

The LFT values of n_8 on u_2 and u_3 are

$$LFT(n_8, u_2) - EST(n_8, u_2) = 93 - 65 = 28 > w_{8,2} = 11, \qquad (3.24)$$

and

$$LFT(n_8, u_3) - EST(n_8, u_3) = 82 - 69 = 13 < w_{10,3} = 14; \qquad (3.25)$$

therefore, a copy can be added to u_2 and a copy cannot be added to u_3 for n_8, as shown in Fig. 3.6.

(4) The same principles as above are used for the remaining tasks. As shown in Fig. 3.7, n_7, n_6, n_4, and n_3 add individual copies in ECUs. Finally, the reliability value of the parallel application is enhanced from 0.95475549 (Fig. 3.5) to 0.97584149 (Fig. 3.7) by using RBFSE to add copies for n_8, n_7, n_6, n_4, and n_3.

The RBFSE algorithm is presented based on the above analysis, as shown in Algorithm 4.

In summary, RBFSE attempts to backward add a new copy for each task to an available ECU that can obtain the maximum reliability value among all available ECUs while meeting precedence constraint and real-time requirement. The RBFSE's time complexity is also $O(|N|^2 \times |U|)$.

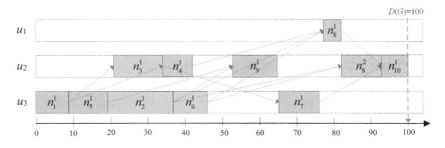

Figure 3.6: A copy is added to u_2 for n_8.

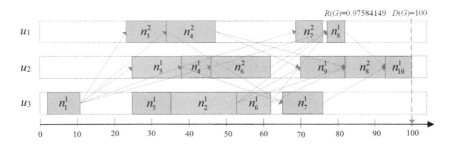

Figure 3.7: Task mapping generated by RBFSE of the motivational parallel application.

3.5.2 RFFSE Algorithm

Similar to the FFSE(non-repeated) algorithm, the RFSE algorithm can be implemented by adding a possible copy in an ECU for each task.

(1) Similar to FFSE algorithm, the forward sequence of tasks is also ordered by the ascending order of the AST values generated by RBFSE. For example, the forward sequence of tasks in Fig. 3.7 is n_1, n_3, n_5, n_4, n_2, n_6, n_7, n_9, n_8, and n_{10}.

(2) First process the entry task n_1. The start time and end time of n_1 are just moved to 0 and 8, respectively, in fixed ECU n_3. No additional copies are added for n_3, n_5, n_4, n_2 until n_6; a new copy for n_6 is added to u_1. That is, n_6 has a total of three copies, as shown in Fig. 3.8.

(3) The tasks n_1, n_3, n_5, n_4, n_2, and n_6 have been processed in Fig. 3.8. Similar to already processed tasks, the remaining tasks n_7, n_9, n_8, and n_{10} will be processed; however, no additional copies are added for n_7, n_9, n_8, and n_{10}, as shown in Fig. 3.9. Finally, the reliability value of the parallel application is increased from 0.97584149 (Fig. 3.7) to 0.97586917 (Fig. 3.9) by using RFSE to add copies for n_6.

Finally, the RFSE algorithm is presented and is shown in Algorithm 5.

Algorithm 4 RBFSE Algorithm

Input: G, $D(G)$, $U = \{u_1, u_2, ..., u_{|U|}\}$, task mapping generated by the previous algorithm

Output: $RT(G)$, $R(G)$, and task mapping

1: Sort *backward_seq* of tasks by the descending order of the AFT values generated by the previous algorithm;
2: **while** (*backward_seq* is not NULL) **do**
3:     $n_i \leftarrow backward\_seq.out()$;
4:     U_{cur} indicates that the ECU set has been assigned;
5:     **for** (each ECU $u_{\text{cur}(i)} \in (U_{\text{cur}})$) **do**
6:         Move the n_i in ECU $u_{\text{cur}(i)}$ by $AFT(n_i)$ and $AST(n_i)$;
7:     **end for**
8:     **for** (each ECU $u_k \in (U - U_{\text{cur}})$) **do**
9:         Calculate $EST(n_i, u_k)$ using Eq. (3.3);
10:        Calculate $LFT(n_i, u_k)$ using Eq. (3.7);
11:        **if** $(((LFT(n_i, u_k) - EST(n_i, u_k)) < w_{i,k})$ **then**
12:            **continue**;
13:        **end if**
14:        Calculate $R(n_i, u_k)$ using Eq. (2.1);
15:     **end for**
16:     Select the $u_{\text{new}(i)}$ with the max $R(G)$ under $R(n_i, u_{\text{new}(i)}) > R(n_i, u_{\text{cur}(i)})$;
17:     **if** ($u_{\text{new}(i)}$ is not NULL) **then**
18:         Add the copy of n_i to ECU $u_{\text{new}(i)}$ by $AFT(n_i)$ and $AST(n_i)$;
19:     **end if**
20:     Calculate $R(n_i)$ using Eq. (3.19);
21: **end while**
22: Calculate $R(G)$ using Eq. (2.2).

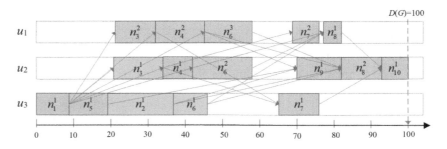

Figure 3.8: A new copy is added to u_1 for n_6 by using RFSE.

3.5.3 Stable Stopping-Based Functional Safety Enhancement

We can perform multiple rounds of forward and backward recovery by using the RBFSE and RFSE algorithms until the reliability values reach stability and fixed values. We explain in Section 3.5.1 that performing non-repetitive BFSE or FFSE

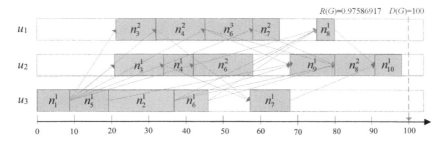

Figure 3.9: Task mapping generated by RFSE of the motivational parallel application.

Algorithm 5 RFSE Algorithm
___

input: $U = \{u_1, u_2, ..., u_{|U|}\}$, G, $D(G)$, task mapping generated by the previous algorithm
output: $R(G)$, $RT(G)$, $rep(G)$, and task mapping
1: Sort $forward\_seq$ by the ascending order of the AST values generated by the previous algorithm;
2: **while** (there are tasks in $forward\_seq$) **do**
3:     $n_i \leftarrow forward\_seq.out()$;
4:     U_{cur} indicates that the ECU set has been assigned;
5:     **for** (each ECU $u_{\text{cur}(i)} \in (U_{\text{cur}})$) **do**
6:         Move the n_i in ECU $u_{\text{cur}(i)}$ by $AST(n_i)$ and $AFT(n_i)$;
7:     **end for**
8:     **for** (each ECU $u_k \in (U - U_{\text{cur}})$) **do**
9:         Calculate $EST(n_i, u_k)$ using Eq. (3.3);
10:        Calculate $LFT(n_i, u_k)$ using Eq. (3.7);
11:        **if** $(((LFT(n_i, u_k) - EST(n_i, u_k)) < w_{i,k})$ **then**
12:            **continue**;
13:        **end if**
14:        Calculate $R(n_i, u_k)$ using Eq. (2.1);
15:     **end for**
16:     Select the $u_{\text{new}(i)}$ with the max $R(G)$ under $R(n_i, u_{\text{new}}(i)) > R(n_i, u_{\text{cur}(i)})$;
17:     **if** ($u_{\text{new}(i)}$ is not NULL) **then**
18:         Add the copy of n_i to ECU $u_{\text{new}(i)}$ by $AST(n_i)$ and $AFT(n_i)$;
19:     **end if**
20:     Calculate $R(n_i)$ using Eq. (3.19);
21: **end while**
22: Calculate $R(G)$ using Eq. (2.2).

---

recovery continuously does not improve functional safety. Fortunately, we can invoke repeated RBFSE and RFSE recovery to enhance functional safety by adding possible new copies to the ECU for each task in each round. The workflow of the SSFSE method is shown in Fig. 3.10.

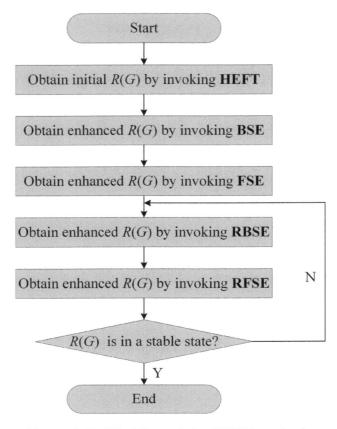

Figure 3.10: Workflow of the SSFSE method.

We explain the process of reliability enhancement of the motivational parallel application as shown in Table 3.5. BFSE, FFSE, RBFSE (1st round), RFFSE (1st round) gradually enhance the reliability value from 0.94771638 to 0.97586917 in rows 3–6. 0.97586917 is a stable and fixed value because RFFSE (1st round), RBFSE (2nd round), and RFFSE (2nd round) generate the same reliability value. Due to the parallel application only includes three ECUs, the second round does not reflect the difference from the first round. The final $RT(G)$ is 98, and the number of copies $rep(G)$ is 16 for the parallel application. The SSFSE method is presented by combining the HEFT and BFSE algorithms with the FFSE, RBFSE, and RFFSE algorithms proposed in Sections 3.4 and 3.5. The SSFSE method can call RBFSE and RFFSE repeatedly until a stable value is reached. Note that RBFSE and RFFSE do not always enhance reliability due to ECU limitations and strict precedence constraint and real-time requirement. Therefore, the SSFSE method can be stopped in a while loop.

The SSFSE method is a recovery through repetition based on forward recovery and backward recovery of the static recovery mechanism, and the method is consistent with the static recovery mechanism in ISO 26262 standard. This standard recommends backward recovery, forward recovery, and recovery through repetition

Table 3.4: Process of reliability enhancement of the motivational parallel application in Fig. 2.3.

Approach	$R(G)$	$RT(G)$	$D(G)$	$rep(G)$	Figure
HEFT	0.94771638	80	100	10	Fig. 3.1
BFSE	0.95294318	92	100	10	Fig. 3.3
FSE	0.95475549	92	100	10	Fig. 3.5
RBFSE (1st round)	0.97584149	98	100	15	Fig. 3.7
RFSE (1st round)	0.97586917	98	100	16	Fig. 3.9
RBFSE (2nd round)	0.97586917	98	100	16	Fig. 3.9
RFSE (2nd round)	0.97586917	98	100	16	Fig. 3.9

Table 3.5: Process of reliability enhancement of the motivational parallel application in Fig. 2.3.

	HEFT	BFSE	FFSE	RBFSE (1st round)	RFFSE (1st round)	RBFSE (2nd round)	RFSE (2nd round)
$R(G)$	0.94771638	0.95294318	0.95475549	0.97584149	0.97586917	0.97586917	0.97586917
$RT(G)$	80	92	92	98	98	98	98
$D(G)$	100	100	100	100	100	100	100
$rep(G)$	10	10	10	15	16	16	16
Figure	Fig. 3.1	Fig. 3.3	Fig. 3.5	Fig. 3.7	Fig. 3.9	Fig. 3.9	Fig. 3.9

as the static recovery mechanism in Section 7.4.12, Part 6 [30]. The SSFSE method combines the above three recovery mechanisms under the precedence constraint and real-time requirement. Therefore, the SSFSE method complies with the functional safety standard of vehicles from a practical perspective. That is, the SSFSE method is a convergence algorithm, which means that the algorithm can stop when the reliability reaches a steady state.

It is undeniable that the SSFSE method has burdens or drawbacks in terms of their ECU as the use of repetition.

(1) If repetition is not used, a task generates a message reception interrupt (MRI) for only a successor task in the ECU that receives its message. For example, task n_8 is a successor task of task n_6 in the motivational parallel application G in Fig. 2.3. After task n_6 is finished, it only needs to send one message to n_8, such that n_6 only generates one MRI to n_8.

(2) If repetition is used, a task only causes multiple MRIs to its one successor task in the ECU that receives its message. For example, after using the SSFSE method, n_6 sends four message MRIs to n_8. The details are shown in Fig. 3.9, where n_6 has three copies of n_6^1, n_6^2, and n_6^3, while n_8 has two copies of n_8^1, and n_8^2. In Fig. 3.9, n_6 generates four MRIs (denoted with red arrows) to n_8.

Many unnecessary MRIs can impose a considerable ECU load since the execution of interrupt service routines (ISRs) and the overhead of interrupt-triggered switching between tasks. These drawbacks are not negligible for safety-critical automotive embedded systems, especially for ECUs running at relatively low clock speeds and with small memory space in the ECUs [13, 45, 46]. Although the SSFSE method enhances the reliability of the parallel application of the automotive embedded systems, it also imposes a considerable ECU load, which will affect the quality of task scheduling.

3.6 EXPERIMENTS FOR FUNCTIONAL SAFETY ENHANCEMENT TECHNIQUE SSFSE

BFSE is selected to compare with the algorithms presented in this chapter. The metrics are reliability $R(G)$, response time $RT(G)$, and copy number $rep(G)$.

The focus of this chapter is on safety prevention using functional safety enhancement techniques during the design phase, while the final implementation results are not determined during the operational phase. Therefore, the automotive parallel applications will be tested by simulation to implement functional safety enhancements in a static recovery manner in this section.

3.6.1 Real-Life Parallel Application

We adopt a real-life parallel application with 31 tasks in Section 2.6.1 and use same parameter values. The lower bound of the parallel application using HEFT algorithms is 630 μs. The response time is changed from 630 μs to 1430 μs with 200 μs increments. The results are shown in Table 3.6–Table 3.8.

(1) The response time generated by all the above algorithms is less than or equal to the corresponding real-time requirement as shown in Table 3.6. Besides that the response time value is equal to the lower bound (i.e., 630 μs), the response time values generated by BFSE and FFSE are less than the real-time requirement as shown in Table 3.6; such results indicate that BFSE and FFSE still leave a larger optimization space for RBFSE, RFFSE, and SSFSE algorithms.

(2) Longer response time could result in higher reliability, and the results generally confirm that maximizing reliability and minimizing response time is a bi-objective optimal problem, as shown in Table 3.7. In addition, the reliability values are classified into three gradients: 1) HEFT; 2) BFSE and FFSE; and 3) RBFSE, RFSE, and SSFSE as shown in Table 3.7 clearly. As expected, SSFSE

Table 3.6: Response time of the real-life parallel application in different real-time requirements.

Deadline	630μs	830μs	1030μs	1230μs	1430μs
HEFT	630μs	630μs	630μs	630μs	630μs
BFSE	630μs	829μs	1010μs	1188μs	1422μs
FFSE	630μs	829μs	1010μs	1188μs	1422μs
RBFSE	630μs	829μs	1028μs	1204μs	1429μs
RFFSE	630μs	829μs	1030μs	1230μs	1430μs
SSFSE	630μs	830μs	1030μs	1230μs	1430μs

Table 3.7: Reliability of the real-life parallel application in different real-time requirements.

Deadline	$2642\mu s$	$3242\mu s$	$3842\mu s$	$4442\mu s$	$5042\mu s$
HEFT	0.94920261	0.94920261	0.94920261	0.94920261	0.94920261
BSE	0.96284388	0.97351212	0.97517143	0.97898692	0.97987526
FSE	0.96284388	0.97524575	0.97517143	0.97898692	0.98013496
RBSE	0.97439984	0.99312022	0.9946906	0.998657	0.998754
RFSE	0.97596001	0.99313219	0.99471573	0.99868679	0.99877729
SSFSE	0.97596002	0.99313283	0.99471575	0.99868684	0.99877734

Table 3.8: Numbers of copies of the real-life parallel application in different real-time constraints.

Deadline	$2642\mu s$	$3242\mu s$	$3842\mu s$	$4442\mu s$	$5042\mu s$
HEFT	31	31	31	31	31
BSE	31	31	31	31	31
FSE	31	31	31	31	31
RBSE	44	57	58	60	60
RFSE	49	70	80	89	88
SSFSE	54	80	91	109	122

method can obtain the maximum reliability value, followed by RFSE, RBFSE, FFSE, BFSE, and HEFT algorithms. The HEFT algorithm has the lowest reliability value of 0.94920261. The reliability value is increased to 0.962−0.980 by backward and forward recovery using BFSE and FFSE algorithms. With the continuous involvement of RBFSE and RFSE (i.e., using SSFSE), the reliability values can reach a steady state in each case. The SSFSE method only increases the reliability by a small amount compared to RFSE because the SSFSE method only invokes RBFSE and RFFSE at most two times. Even so, functional safety can be maximized by simply increasing the slack response time.

ISO 26262 provides the duration/probability of exposure levels as shown in Table 3.9. E1 means very low probability with reliability requirement > 0.99, E2 means low probability with reliability requirement 0.99, E3 means medium probability with reliability requirement > 0.9 and < 0.99, and E4 means high probability with reliability requirement <= 0.9. In Table 3.7, the maximum reliability value using BFSE is 0.97987526 (E3), while using the SSFSE method reaches 0.99877734 (E1). That is, the exposure level decreases from E3 to E1 using the SSFSE method. This reduction in exposure (i.e., reliability enhancement) results in a higher level of functional safety assurance.

Although the reliability enhancement of the SSFSE method for RBFSE and RFSE in Table 3.7 is small, the results are still useful for safety-sensitive automotive embedded systems. The exposure probability categories in ISO 26262 shown in Table 3.9 are informative, not prescriptive, and leave a great deal of discretion to those building each component system and ultimately to vehicle manufacturers and suppliers [24]. For example, ISO 26262 does not specify an exposure probability E1, which corresponds to a reliability target of at least more than 0.99, as shown in Table 3.9. Hence, different vehicle manufacturers and suppliers can choose different reliability requirements depending on their products and market orientations, as long as these values fall under the same exposure. For example, a reliability requirement of 0.99877730 for E3 is feasible because this value exceeds 0.99, and ISO 26262 requires a high level of accuracy for reliability value. When the real-time requirement is 1430 μs as shown in Table 3.7, the SSFSE method reduces to E1, while RBFSE and RFSE only reduce to E2. This result fully demonstrates the advantage of the SSFSE method.

Table 3.9: Classification of the duration/probability of exposure in ISO 26262 [28, 30].

Exposure level	E1 Very low probability	E2 Low probability	E3 Medium probability	E4 High probability
Probability of exposure	Not specified	<1%	[1%, 10%]	>10%
Reliability requirement	At least exceeds 0.99	0.99	>0.9	<=0.9

(3) Table 3.8 clearly displays that copies generated by HEFT, BFSE, and FFSE are fixed at 31 because the primary backup repetition cannot be implemented. The number of copies generated by the RBFSE, RFFSE, and SSFSE methods gradually increases. For instance, when the real-time requirement is 630 μs, the number of copies generated by RBFSE, RFSE, and SSFSE is 44, 49, and 54, respectively. As the reliability value increases with the real-time requirement, the number of copies required increases with the primary backup repetition. For instance, when the real-time requirement is 1430 μs, the number of copies generated by RBFSE, RFSE, and SSFSE is 60, 88, and 122, respectively. Even though primary backup repetition is a practical fault tolerance measure to improve functional safety, it also has some disadvantages. Namely, RBFSE, RFSE, and SSFSE, especially the SSFSE method increases the burden on the ECUs due to extra copies and affect the control efficiency of the ECUs due to the time-consuming extra code. These problems should be considered as disadvantages of fault tolerance measures.

(4) In each case, the time taken to calculate the results using the SSFSE method is very short, within 1s. This is because the SSFSE method only calls RBFSE and RFSE at most twice.

Table 3.10: Response time of the synthetic parallel application in different real-time requirements.

Deadline	$2642\,\mu s$	$3242\,\mu s$	$3842\,\mu s$	$4442\,\mu s$	$5042\,\mu s$
HEFT	$2642\,\mu s$	$2642\,\mu s$	$2642\,\mu s$	$2642\,\mu s$	$2642\,\mu s$
BFSE	$2642\,\mu s$	$3240\,\mu s$	$3842\,\mu s$	$4442\,\mu s$	$5042\,\mu s$
FSE	$2642\,\mu s$	$3238\,\mu s$	$3839\,\mu s$	$4440\,\mu s$	$5042\,\mu s$
RBFSE	$2642\,\mu s$	$3242\,\mu s$	$3842\,\mu s$	$4442\,\mu s$	$5042\,\mu s$
RFFSE	$2642\,\mu s$	$3242\,\mu s$	$3842\,\mu s$	$4442\,\mu s$	$5042\,\mu s$
SSFSE	$2642\,\mu s$	$3242\,\mu s$	$3842\,\mu s$	$4442\,\mu s$	$5042\,\mu s$

Table 3.11: Reliability of the synthetic parallel application in different real-time requirements.

Deadline	$2642\,\mu s$	$3242\,\mu s$	$3842\,\mu s$	$4442\,\mu s$	$5042\,\mu s$
HEFT	0.88677322	0.88677322	0.88677322	0.88677322	0.88677322
BFSE	0.92139357	0.92689035	0.92956365	0.93075425	0.93212067
FFSE	0.92139357	0.92839499	0.93107262	0.93226516	0.93308311
RBFSE	0.95441984	0.96679232	0.96948084	0.97002576	0.97817453
RFFSE	0.95443768	0.96681756	0.96950724	0.97004826	0.9781965
SSFSE	0.9544377	0.96681759	0.96950726	0.97004828	0.97819651

3.6.2 Synthetic Parallel Application

Besides using a real-life parallel application of an embedded system to confirm the advantages of the presented algorithm, an extra synthetic application with 100 tasks is used to analyze the results. The synthetic application has the same parameter values as the real-life parallel application. We also use a task graph generator to generate a randomly parallel application [1]. The parameter settings for this experiment are the same as in Section 2.6.2.

The lower bound for the parallel application generated by HEFT is 2642 μs. As the real-time requirement must be larger than or equal to the lower bound explained in Section 3.3.1, the real-time requirement is changed from 2642 μs to 5042 μs with 600 μs increments. The results are shown in Table 3.10–Table 3.12.

(1) Similar to Table 3.6, Table 3.10 shows that the response time values generated by all the algorithms are also less than or equal to the corresponding real-time requirement. In summary, Tables 3.6 and 3.10 have the same pattern.

(2) Similar to Table 3.7, Table 3.11 also explicitly divides the reliability values into three gradients: 1) HEFT algorithm; 2) BFSE and FFSE algorithms; and 3) RBFSE, RFSE, and SSFSE algorithms. The HEFT algorithm has the lowest reliability value of 0.88677322, which is much lower than 0.94920261 as shown in table 3.7. The reason is that the reliability value of a parallel application is the product of all tasks, so that the reliability value of a parallel application

Table 3.12: Numbers of copies of the synthetic parallel application in different real-time requirements.

Deadline	$2642\,\mu s$	$3242\,\mu s$	$3842\,\mu s$	$4442\,\mu s$	$5042\,\mu s$
HEFT	100	100	100	100	100
BFSE	100	100	100	100	100
FSE	100	100	100	100	100
RBFSE	153	165	166	168	167
RFSE	185	210	213	218	218
SSFSE	248	305	345	378	392

generated by 100 tasks is naturally lower than the reliability value generated by 31 tasks (the real-life parallel application contains 31 tasks) . In general, Tables 3.7 and 3.11 show the same pattern for all algorithms.

(3) Table 3.12 presents the same pattern as Table 3.8. By synthesizing the data in Tables 3.8 and 3.12, the following facts are confirmed. 1) HEFT, BFSE and FFSE have a fixed number of copies $|N|$ because there is no primary backup repetition, and $|N|$ represents the task set size for the parallel application. 2) RBFSE generates a number of copies between $|N|$ and $2 \times |N|$ since each task is repeated at most once. 3) The number of copies generated by RFSE is between $|N|$ and $3 \times |N|$ because each task is repeated at most once on the basis of RBFSE. 4) The copies generated by SSFSE is theoretically between $|N|$ and $|U| \times |N|$, where $|U|$ denotes the size of the ECU set, but it is about $4 \times |N|$ in practice.

(4) In each case, the time taken to calculate the results using the SSFSE method is still within 1s, since the SSFSE method only calls RBFSE and RFSE up to 5 times for an automotive parallel application with 100 tasks.

3.7 CONCLUDING REMARKS

We present SSFSE method for a parallel application of an embedded system to enhance functional safety in this chapter. SSFSE method combines HEFT, BFSE algorithms and FFSE, RBFSE, and RFSE algorithms presented in this chapter. SSFSE method enhances functional safety by using a stable stopping method on the basis of the forward and backward recovery through primary-backup repetition. SSFSE method can bring the exposure level down from E3 to E1 toward a higher level of safety assurance. SSFSE method is actually a repeated recovery based on backward recovery and forward recovery of the static recovery mechanism as indicated in the second edition of ISO 26262. It is believed that the stable stopping method can be used as a guideline in the functional safety design of parallel applications of embedded systems.

CHAPTER 4

Functional Safety Validation

F UNCTIONAL safety requirements of automotive embedded systems include response time and reliability requirements. These two requirements must be met at the same time to ensure the automotive functional safety requirements. However, increasing reliability comes with extra response time overhead. This chapter addresses a method to get minimum response time possible while meeting reliability requirement. In recent years, a practical reliability requirement validation method was come up with that distributed the reliability requirement of the application to every single task to preassign reliability values to unallocated tasks. This chapter introduces two algorithms, namely non-fault tolerant reliability preassignment based on geometric mean (GMNRA) and fault tolerant reliability preassignment based on geometric mean (GMFRA) , where reliability values based on geometric mean are preassigned to unassigned tasks. Since geometric mean can make the preassigned reliability values of unassigned tasks to the central tendency, it distributes the reliability requirements in a more balanced way. At the end of this chapter, massive experiments are given to verify the efficiency of GMNRA and GMFRA to decrease the response time.

4.1 INTRODUCTION

With the improvement of vehicle safety requirements, many end-to-end computing functions have been introduced and integrated into vehicle systems. For example, the brake-by-wire function, a typical automotive function, is released when it received collected data by one sensor and is executed in five ECUs, transmitted on two CAN buses, and achieved after sending the executive action to two actuators [48]. Given that the mentioned process is directed and acyclic, DAG was presented to describe a distributed automotive application [25, 61, 71, 72]. In the DAG model, the nodes signify the tasks and the edges signify the messages between tasks' communication [25, 61, 71, 72].

Automotive functional safety requirements consist of response time and reliability requirements according to ISO 26262 [28]. From a functional point of view, reliability can be regarded as the possibility of the application surviving during a specific period. According to ISO 26262, random hardware failures (i.e., transient failures) occur in an unpredictable manner throughout the lifecycle of a hardware component, but follow a probability distribution [28]. In the above standard, reliability is inversely

DOI: 10.1201/9781003391517-4

represented as exposure (i.e., reliability = 1−exposure) referring to the relative expected frequency of operational conditions, where injurious and hazardous events may occur [28]. Exposure is divided into five levels (i.e., E0, E1, E2, E3, and E4) [28, 77].

Both reliability and real-time requirements must be fulfilled simultaneously to ensure automotive functional safety requirements. However, reliability and response time are hard to be met simultaneously in reality because raising reliability intuitively raises the response time of a distributed application based on the DAG model [20, 61]. Since minimizing response time and maximizing reliability are conflicting, ensuring functional safety requirements is a bicriteria optima problem [20, 61, 62]. In view of this sitution, we intend to present an approach to search for the solution to the minimum response time while not violating reliability requirement.

The development lifecycle of the parallel application of automotive embedded systems usually includes the analysis, design, implementation, and testing phases [61]. Our job attempts to present efficient reliability requirement validation ways for an automotive application throughout the design phase. Based on this purpose, we introduce the concept of geometric mean, implement pre-assigned reliability values and take the response time of the application as the optimization target. The principal techniques of this chapter are summarized as follows:

(1) We propose a geometric mean-based non-fault tolerant reliability preassignment (GMNRA) algorithm, where geometric mean-based reliability values are preassigned to unallocated tasks.

(2) We propose a geometric mean-based fault tolerant reliability preassignment (GMFRA) algorithm, where geometric mean-based reliability values are preassigned to unallocated tasks.

(3) Geometric mean shows its advantage to concentrate the preassigned reliability values of unallocated tasks. Experimental results represent that the GMNRA and GMFRA algorithms are effectively in reducing response times compared to state-of-the-art algorithms.

4.2 RELATED WORK

Safety validation is a comprehensive and sufficient judgment of whether the system achieves the safety requirement, which consists of non-fault tolerant validation methods and fault tolerant validation methods. In this chapter, we primarily review those studies about the reliability requirement validation of an application based on DAG model organized by (1) non-fault tolerant functional safety validation and (2) fault tolerant functional safety validation.

(1) non-fault tolerant reliability requirement validation. Transient failures that follow the Poisson distribution have been mentioned at high frequency observed in many studies [20, 61, 62, 75, 76]. However, the high reliability raises the response time of a parallel automotive embedded application base on the DAG model, and optimizing response time and reliability is considered to be a typical bi-objective optimization or Pareto optimization problem [17, 18, 20, 61]. Reference [17] proposed a method to solve the bi-objective optimization problem between reliability and response time by merging the bi-objective of response time and reliability into a single objective application for their joint optimization. Reference [17] presented a method to obtain better reliability but a longer response time than that in Reference [20]. Reference [18] proposed an approach that implemented the bi-objective optimization for reliability maximization and response time minimization. However, References [17, 18, 20] do not consider ensuring the reliability requirement of the application. Reference [61] presented the MRCRG method and addressed the reliability requirement assurance problem by preassigning the maximum reliability value to each unallocated task.

(2) Fault tolerant reliability requirement validation. There are three major fault tolerant reliability requirement validation methods according to the literature. The MaxRe method presented in Reference [76] devotes to the resource cost minimization of an application while ensuring its reliability requirement. MaxRe distributes the reliability requirement to each task by setting all tasks with an equal reliability requirement at an average level. The RR method presented in Reference [75] improves the MaxRe approach by considering the previous assignments of tasks when assigning the current task. Reference [62] studied the MaxRe and RR in detail and presented the improved HRRM algorithm by preassigning the upper reliability values to unassigned tasks.

4.3 MODELS

Tables 4.1 and 4.2 illustrate the abbreviations and notations that are used in this chapter.

Table 4.1: Abbreviations in this chapter.

Abbreviation	Definition
WCTT	Worst Case Execution Time
ASW	Application Software
RTE	Run-Time Environment
BSW	Basic Software
AUTOSAR	Automotive Open System Architecture
MRTRR	Minimizing Response Time under Reliability Requirement
GMNRA	Geometric Mean-based non-fault tolerant Reliability Preassignment
GMFRA	Geometric Mean-based Fault tolerant Reliability Preassignment
HRRTM	Heuristic Replication for Response Time Minimization

4.3.1 System Architecture

We use the same embedded system architecture and parallel application model as in Section 2.3.1. The entry task of the parallel application can only be executed by specific ECUs connecting sensors, and the exit task of the parallel application can only be executed by specific ECUs connecting actuators in this architecture.

Table 4.2: Notations in this chapter.

Notation	Definition
num_i	Number of copies of the task n_i
$n_{s(y)}$	yth assigned task
n_i^β	βth copy of task n_i
$R_{\max}(n_i)$	Maximum reliability of task n_i under non-fault tolerance
$R_{\mathrm{fmax}}(n_i)$	Maximum reliability of task n_i under fault tolerance
$R_{\mathrm{up}}(n_i)$	Upper reliability value of task n_i
$GM(G)$	Geometric mean of the parallel application G
$R_{\mathrm{pre}}(n_{s(z)})$	Preassigned reliability value of task $n_{s(z)}$
$R_{\mathrm{req}}(n_{s(y)})$	Reliability requirement of task $n_{s(y)}$
$GM'(G)$	Geometric mean of the parallel application G by using GMFRA
$R'_{\mathrm{pre}}(n_{s(z)})$	Preassigned reliability value of task $n_{s(z)}$ by using GMFRA
$R'_{\mathrm{req}}(n_{s(y)})$	Reliability requirement of task $n_{s(y)}$ by using GMFRA

AUTOSAR defines communication processes within and among ECUs in CAN bus [42]. AUTOSAR consists of three levels, application software(ASW), run-time environment (RTE), and basic software (BSW). If two tasks belong to the same ECU, they communicate with each other through the RTE communication services (i.e., shared memory mechanism). Otherwise, their intercommunication needs to span multiple layers of the AUTOSAR specification and eventually use the network to complete the intercommunication (i.e., the message passing mechanism) [42]. In a 32-bit 400 MHz shared memory mechanism, the worst-case transfer time (WCTT) of a packet in memory is about $0.01\mu s$ for a 128-bit extended CAN frame. In a message passing mechanism with a CAN speed of 500 Kbit/s, the WCTT of a packet in the bus is approximately 256 μs for a 128-bit extended CAN. Given that the speed in shared memory is much faster than the speed of message passing, the communication time between tasks assigned to the same ECU is negligible. Hence, the communication time can be set to 0 for n_i and n_j assigned to the same ECU. The scheduling can be either preemptive or non-preemptive on the basis of the AUTOSAR standard. We only take non-preemptive scheduling for ECUs into consideration in this chapter for simplicity.

4.3.2 Reliability Model

In this chapter, non-fault tolerant and fault tolerant mechanisms are described and a reliability model is established based on the above mechanisms.

(1) No tasks can be replicated under the non-fault tolerance mechanism. Hence, the reliability value of n_i running on u_k in its WCET is the same as Eq. (2.1).

The maximum and minimum reliability values of each task are respectively calculated by

$$R_{\max}(n_i) = \max_{u_k \in U} R(n_i, u_k). \tag{4.1}$$

and

$$R_{\min}(n_i) = \min_{u_k \in U} R(n_i, u_k), \qquad (4.2)$$

(2) Under the mechanism of fault tolerance, since the parallel application is implemented with software, a task has up to $|U|$ copies (including primary and backups). Therefore, the maximum and minimum reliability values of each task are respectively calculated by

$$R_{\mathrm{fmax}}(n_i) = 1 - \prod_{k=1}^{|U|} (1 - R(n_i, u_k)). \qquad (4.3)$$

and

$$R_{\min}(n_i) = \min_{u_k \in U} R(n_i, u_k), \qquad (4.4)$$

Assuming that n_i has num_i copies, its reliability value is calculated by

$$R(n_i) = 1 - \prod_{l=1}^{num_i} (1 - R(n_i, u_{k_l})), \qquad (4.5)$$

where u_{k_l} represents the assigned ECU of the l-th copy of n_i.

Instead of differential transmission specifications at the physical layer, the CAN protocol provides sub-packet authentication rules at the data link layer using CRC and ACK segments. Such a measure guarantees a CAN packet failure rate of 10^{-9} [15] in a normal environment, which is much lower than the ECU failure rate of 10^{-6} [33]. Therefore, in this chapter, only ECU failures are considered and communication failures are ignored. Hence, the reliability of the parallel application G is the product of the reliability values of all tasks, that is,

$$R(G) = \prod_{n_i \in N} R(n_i). \qquad (4.6)$$

ISO 26262 provides the duration/probability of exposure [28] as shown in Table 3.9. For given exposure levels, we can deduce the corresponding reliability requirement. For instance, the probability of exposure E2 is less than 1% of the average operating time. This means that the minimum probability of a hazardous event is close to 0.01. In order to ensure safety, the actual reliability must be equal to or greater than $1 - 0.01 = 0.99$, which is considered to be the reliability requirement in this case. Similarly, the reliability requirements for other exposures can be calculated according to the above rules. The reliability requirements of the exposures are shown in Table 3.9.

4.3.3 Problem Statement

The purpose of this chapter is to obtain the task assignments in ECUs to reduce the response time of the motivational parallel application G,

$$RT(G) = AFT(n_{\mathrm{exit}}), \qquad (4.7)$$

and ensure its reliability requirement at the same time,

$$R(G) = \prod_{n_i \in N} R(n_i) \geq R_{\text{req}}(G). \tag{4.8}$$

$AFT(n_{\text{exit}})$ denotes the AFT of the exit task. Because scheduling tasks with quality of service (QoS) requirements on multiple ECUs is an NP-hard optimization problem [59] and the long development lifecycle of automotive embedded systems, we try to present a heuristic method to solve this problem.

4.4 NON-FAULT TOLERANT FUNCTIONAL SAFETY VALIDATION

4.4.1 Non-Fault Tolerant Reliability Requirement Assessment

Under the mechanism of non-fault tolerance, considering that the reliability of the parallel application G is the product of the reliability values of all tasks, (Eq. (4.6)), the maximum and minimum reliability values of the parallel application G are calculated by

$$R_{\max}(G) = \prod_{n_i \in N} R_{\max}(n_i). \tag{4.9}$$

and

$$R_{\min}(G) = \prod_{n_i \in N} R_{\min}(n_i), \tag{4.10}$$

Before functional safety can be validated, $R_{\text{req}}(G)$ must be evaluated. First, $R_{\text{req}}(G)$ must be greater than or equal to $R_{\min}(G)$; otherwise, $R_{\text{req}}(G)$ will not always be satisfied. Also, $R_{\text{req}}(G)$ must be less than or equal to $R_{\max}(G)$; otherwise, $R_{\text{req}}(G)$ is absolutely not guaranteed. Therefore, $R_{\text{req}}(G)$ must be in the range of $R_{\min}(G)$ and $R_{\max}(G)$, that is,

$$R_{\min}(G) \leqslant R_{\text{req}}(G) \leqslant R_{\max}(G). \tag{4.11}$$

4.4.2 Existing Non-Fault Tolerant Functional Safety Validation Algorithms

As described in Section 4.2, the MRCRG algorithm introduced in Reference [61] transfers the reliability requirement of the parallel application to each task. In this chapter, this approach is called minimizing response time and reliability requirements (MRRTR). The main idea of MRTRR reliability requirement validation is as follows.

Assuming that the task to be assigned is $n_{s(y)}$, which represents the y-th assigned task, $\{n_{s(1)}, n_{s(2)}, ..., n_{s(y-1)}\}$ denotes the task set where the tasks have been assigned, and $\{n_{s(y+1)}, n_{s(y+2)}, ..., n_{s(|N|)}\}$ represents the task set where the tasks have not been assigned. Initially, all tasks of the parallel application are unassigned. The task priority is determined by Eq. (2.6) according to the descending order of upward rank value used in Reference [58].

To ensure the reliability requirement of the parallel application at each task assignment, MRTRR presupposes that each unassigned task in $\{n_{s(z+1)}, n_{s(z+2)}, ..., n_{s(|N|)}\}$ is preassigned with the maximum reliability value when assigning $n_{s(y)}$.

Hence, the reliability of the parallel application G is calculated by

$$R(G) = \prod_{x=1}^{y-1} R(n_{s(x)}) \times R(n_{s(y)}) \times \prod_{z=y+1}^{|N|} R_{\max}(n_{s(z)}). \tag{4.12}$$

Next, the actual reliability $R(G)$ must be equal to or larger than $R_{\text{req}}(G)$ based on the problem statement. In this situation, we have

$$\prod_{x=1}^{y-1} R(n_{s(x)}) \times R(n_{s(y)}) \times \prod_{z=y+1}^{|N|} R_{\max}(n_{s(z)}) \geqslant R_{\text{req}}(G). \tag{4.13}$$

Hence, the actual reliability value of task $n_{s(y)}$ must meet the following constraint:

$$R(n_{s(y)}) \geqslant \frac{R_{\text{req}}(G)}{\prod_{x=1}^{y-1} R(n_{s(x)}) \times \prod_{z=y+1}^{|N|} R_{\max}(n_{s(z)})}. \tag{4.14}$$

Then, MRTRR defines the reliability requirement of task $n_{s(y)}$ be

$$R_{\text{req}}(n_{s(y)}) = \frac{R_{\text{req}}(G)}{\prod_{x=1}^{y-1} R(n_{s(x)}) \times \prod_{z=y+1}^{|N|} R_{\max}(n_{s(z)})}. \tag{4.15}$$

After the above processing, the reliability requirement of the parallel application is transferred to each task. As long as each task meets its reliability requirement of

$$R(n_{s(y)}) \geqslant R_{\text{req}}(n_{s(y)}), \tag{4.16}$$

then the reliability requirement of the parallel application can be met.

Similar to the HEFT [58], MRTRR assigns n_i (i.e., $n_{s(y)}$) to the ECU with the minimum EFT by using the insertion-based scheduling method as well , which is a partially optimal heuristic method. The EFT of n_i on u_k is calculated by

$$EFT(n_i, u_k) = EST(n_i, u_k) + w_{i,k}, \tag{4.17}$$

where $EST(n_i, u_k)$ represents the EST of n_i on u_k and is calculated by

$$\begin{cases} EST(n_{\text{entry}}, u_k) = 0, \\ EST(n_i, u_k) = \max \left\{ \begin{matrix} avail[k], \\ \max_{n_h \in pred(n_i)} \{AFT(n_h) + c'_{h,i}\} \end{matrix} \right\}, \end{cases} \tag{4.18}$$

where $avail[k]$ is the earliest available time when u_k is ready to execute the task, $AFT(n_h)$ is the AFT of task n_h, and $c'_{h,i}$ denotes the WCRT between n_h and n_i. If n_h and n_i are assigned to the same ECU, then $c'_{h,i} = 0$; otherwise, $c'_{h,i} = c_{h,i}$.

Considering that the reliability requirement of each task has been determined in Eq. (4.15), MRTRR minimizes the AFT of each task by traversing all ECUs while ensuring their reliability requirement. That is, the assigned $u_{\mathrm{pr}(i)}$ for n_i is calculated by

$$AFT(n_i) = EFT(n_i, u_{\mathrm{pr}(i)}) = \min_{u_k \in U, R(n_i, u_k) \geqslant R_{\mathrm{req}}(n_i)} \{EFT(n_i, u_k)\}. \qquad (4.19)$$

In this way, the low complexity MRTRR algorithm is implemented.

Even though the MRTRR algorithm can achieve its intended purpose, preassigning maximum reliability value for each unassigned task results in an unbalanced reliability distribution for all tasks, which leads to a limited reduction in response time. We adopt the following example to explain this.

4.4.3 Example of the MRTRR Algorithm

We use the example of the motivational parallel application in Fig. 2.3 to explain the algorithm. We assume that the failure rates of u_1, u_2, and u_3 are $\lambda_1 = 0.0002$, $\lambda_2 = 0.0005$, and $\lambda_3 = 0.0009$, respectively. The minimum reliability value of the parallel application is 0.879238, and the maximum reliability value is 0.975602 (calculated by Eqs. (4.9) and (4.10)). The reliability requirement of the parallel application is set to $R_{\mathrm{req}}(G) = 0.95$. Note that the above values do not represent real deployments and are only meant to clearly explain the example.

Table 4.3 shows the task assignment for an incentivized parallel application using the MRTRR algorithm. Each row shows the selected ECU (indicated by a box) and corresponding EFT. In order to provide the reader with a full understanding of the reliability requirement assurance process, we provide some of the key details below.

Table 4.3: Task assignment generated by MRTRR of the motivational parallel application.

n_i	$R_{\mathrm{req}}(n_i)$	$R(n_i, u_1)$	$EFT(n_i, u_1)$	$R(n_i, u_2)$	$EFT(n_i, u_2)$	$R(n_i, u_3)$	$EFT(n_i, u_3)$	$R(n_i)$
n_1	0.991933	0.997204	14	0.992032	16	**0.991933**	**9**	0.991933
n_3	0.983045	0.997802	32	0.993521	34	**0.983045**	**28**	0.983045
n_4	0.991047	0.997403	31	**0.996008**	**26**	0.984816	-	0.996008
n_2	0.992435	**0.997403**	**40**	0.990545	-	0.983931	-	0.997403
n_5	0.992634	0.997603	52	**0.993521**	**39**	0.991040	-	0.993521
n_6	0.996513	**0.997403**	**53**	0.992032	-	0.991933	-	0.997403
n_9	0.995517	**0.996406**	**71**	0.994018	-	0.982161	-	0.996406
n_7	0.997709	**0.998601**	**78**	0.992528	-	0.990149	-	0.998601
n_8	0.998108	**0.999000**	**83**	0.994515	-	0.987479	-	0.999000
n_{10}	0.995616	0.995809	104	**0.996506**	**102**	0.985703	-	0.996506
$R(G) = 0.950848 > R_{\mathrm{req}}(G) = 0.95$, $RT(G) = 102$								

(1) Given that the task priority is determined by Eq. (2.6) as shown in Table 2.4, the task priority is n_1, n_3, n_4, n_2, n_5, n_6, n_9, n_7, n_8, n_{10}.

(2) In this case, MRTRR first considers n_1 because n_1 has the first priority. The reliability requirement of n_1 is 0.991933 calculated by using Eq. (4.15). The reliability values of n_1 on u_1, u_2, and u_3 are 0.997204, 0.992032, and 0.991933

calculated by Eq. (2.1). These reliability values all exceed n_1's reliability requirement of 0.991933. The EFTs calculated by Eq. (4.17) of n_1 on u_1, u_2, and u_3 are 14, 16, and 9. In this case, MRTRR selects u_3 because it has the minimum EFT of 9 (denoted with boxed). Accordingly, task mapping of u_1 is on u_3 (0−9) as shown in Fig. 4.1.

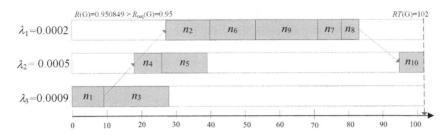

Figure 4.1: Task mapping generated by MRTRR of the motivational parallel application.

(3) MRTRR second considers n_3 because n_3 has the second priority. The task n_3 uses the same pattern with n_1, and its task mapping is on u_3 (9−28) as shown in Fig. 4.1.

(4) Next, MRTRR considers n_4 according to the task priority. The reliability requirement of n_4 is 0.991047 calculated by Eq. (4.15). The reliability values of n_1 on u_1, u_2, and u_3 are 0.997403, 0.996008, and 0.984816 calculated by Eq. (2.1). However, 0.984816 is less than the reliability requirement of n_4 of 0.991047. Therefore, only u_1 and u_2 can be computed with EFT values 31 and 26, respectively, and u_3 is denoted by "−". In this case, MRRTR chooses ECU u_2 because it has a minimum EFT of 26 (indicated by boxed). Accordingly, the task mapping of n_4 on u_2 (18 − 26) is shown in Fig. 4.1.

(5) The remaining tasks adopt the same pattern as n_1, n_3, and n_4. Finally, the response time and actual reliability value of the parallel application G are 102 and 0.950848, respectively. The final task mapping of the motivational parallel application is shown in Fig. 4.1.

This example illustrates that the use of the MRTRR algorithm ensures that the actual reliability of the parallel application is greater than or equal to the reliability requirement. However, as shown in Table 4.3, the high priority tasks $n_1 - n_5$ are scheduled with a low reliability requirement of 0.983045 − 0.992634, while the low-priority tasks $n_6 - n_{10}$ are scheduled with a high reliability requirement of 0.995517 − 0.998108. As a result, the high-priority tasks consume insufficient reliability, which results in insufficient space for the low-priority tasks to select ECUs with less EFT. As an example, for n_6, n_9, n_7, and n_8, there is only one ECU selection. Therefore, low-priority tasks may lengthen the response time of the parallel application. This example essentially reflects that the preassigned reliability values in MRTRR take a greedy method to mapping high-priority tasks, and thus may compromise the overall system responsiveness due to unreliable low-priority tasks. The reliability distribution of all tasks in MRTRR is unbalanced, resulting in a limited reduction in response time of the parallel applications.

4.4.4 Use of Geometric Mean under Non-Fault Tolerance

Considering the ineffective use of MRTRR in reliability requirement validation, we propose an effective approach to reliability requirement validation. If we can decrease the imbalance of preassigned reliability values to unassigned tasks, then we can obtain faster response times.

In mathematics, the mean represents a concentrated trend in a set of numbers that allows for a more balanced assignment of values. There are three general types of means: the geometric mean, the arithmetic mean, and the harmonic mean.

The Geometric mean is the $|N|$-th root of the product of $|N|$ values. For a set of numbers $t_1, t_2, ..., t_{|N|}$, the geometric mean is

$$GM = \sqrt[|N|]{t_1 \times t_2 \times ... \times t_{|N|}}. \tag{4.20}$$

The arithmetic mean is the sum of the $|N|$ values divided by the $|N|$. For a set of numbers $t_1, t_2, ..., t_{|N|}$, the arithmetic mean is

$$AM = \frac{t_1 + t_1 + ... + t_{|N|}}{|N|}. \tag{4.21}$$

The harmonic mean is $|N|$ divided the sum of the reciprocals $|N|$ values. For a set of numbers $t_1, t_2, ..., t_{|N|}$, the harmonic mean is

$$HM = \frac{|N|}{\frac{1}{t_1} + \frac{1}{t_2} + ... + \frac{1}{t_{|N|}}}. \tag{4.22}$$

These three means can indicate the central tendency of a set of numbers. We use the geometric mean for the following main reasons:

(1) The reliability of the parallel application is the product of the reliability values of all tasks. It is the geometric mean that uses the product of its values to represent the central tendency, while the arithmetic means and harmonics mean to use the sum of their values.

(2) Both the arithmetic mean and the harmonic mean are susceptible to extreme values at both ends. The larger the upper limit value, the greater the average upward deviation from the concentrated trend. Conversely, the lower the value, the greater the average downward deviation from the concentrated trend. The geometric mean has the advantage that it is less susceptible to extreme values.

If we find concentrated trends in preassigned reliability values, then we can reduce the imbalance. Hence, we can use the geometric mean to find the concentrated trend. We devise the geometric mean as follows.

Since the reliability requirement of the parallel application G is $R_{req}(G)$, the upper reliability value of each task is calculated by

$$R_{up}(n_i) = \sqrt[|N|]{R_{req}(G)}. \tag{4.23}$$

As long as the actual reliability of each task is greater than or equal to $R_{up}(n_i)$, the reliability requirement of the parallel application is guaranteed because the reliability value is less than 1. However, this approach is also unbalanced. Hence, we present the geometric mean-based reliability preassignment method as follows.

(1) We first define the geometric mean of the parallel application. Considering that the minimum, maximum, and upper reliability values for each task are known, the geometric mean in this chapter is defined as the $|N|$-th root of the product of $|N|$ tasks, that is,

$$
\begin{aligned}
t_1 &= \frac{R_{up}(n_1)}{R_{min}(n_1) \times R_{max}(n_1)}, \\
t_2 &= \frac{R_{up}(n_2)}{R_{min}(n_2) \times R_{max}(n_2)}, \\
&\quad \dots, \\
t_{|N|} &= \frac{R_{up}(n_{|N|})}{R_{min}(n_{|N|}) \times R_{max}(n_{|N|})}.
\end{aligned}
\tag{4.24}
$$

Hence, the geometric mean of the parallel application is

$$
\begin{aligned}
GM(G) &= \sqrt[|N|]{t_1 \times t_2 \times, \dots, \times t_{|N|}} \\
&= \sqrt[|N|]{\prod_{i=1}^{|N|} \frac{R_{up}(n_i)}{R_{min}(n_i) \times R_{max}(n_i)}} \\
&= \sqrt[|N|]{\frac{R_{req}(G)}{R_{min}(G) \times R_{max}(G)}}.
\end{aligned}
\tag{4.25}
$$

(2) We set the preassigned reliability value of $n_{s(z)}$ be

$$
\begin{aligned}
R_{pre}(n_{s(z)}) &= GM(G) \times R_{min}(n_{s(z)}) \times R_{max}(n_{s(z)}) \\
&= \sqrt[|N|]{\frac{R_{req}(G)}{R_{min}(G) \times R_{max}(G)}} \times R_{min}(n_{s(z)}) \times R_{max}(n_{s(z)}),
\end{aligned}
\tag{4.26}
$$

by multiplying $GM(G)$ by the coefficient $R_{min}(n_{s(z)}) \times R_{max}(n_{s(z)})$. Such an operation ensures that each task always finds an ECU assignment to guarantee the reliability requirement of the parallel application.

After the above processing, we set the preassigned reliability values for the unassigned tasks to a concentrated trend, which allows for a more balanced assignment of reliability requirement. In other words, the geometric mean will have the effect of assigning nominal values instead of selecting extreme values. The reliability requirement when assigning $n_{s(y)}$ is calculated by

$$R_{\text{req}}(n_{s(y)}) = \frac{R_{\text{req}}(G)}{\prod\limits_{x=1}^{y-1} R(n_{s(x)}) \times \prod\limits_{z=y+1}^{|N|} R_{\text{pre}}(n_{s(z)})}. \tag{4.27}$$

Furthermore, assuming that the minimum reliability of $n_{s(y)}$ is $R_{\min}(n_{s(y)})$, $R_{\text{req}}(n_{s(y)})$ must be updated to

$$R_{\text{req}}(n_{s(y)}) = \max\left\{ R_{\min}(n_{s(y)}), R_{\text{req}}(n_{s(y)}) \right\}. \tag{4.28}$$

Assuming that the maximum reliability of $n_{s(y)}$ is $R_{\max}(n_{s(y)})$, $R_{\text{req}}(n_{s(y)})$ must be further updated to

$$R_{\text{req}}(n_{s(y)}) = \min\left\{ R_{\max}(n_{s(y)}), R_{\text{req}}(n_{s(y)}) \right\}. \tag{4.29}$$

For any task $n_{s(y)}$, if and only if

$$R\left(n_{s(y)}\right) \geqslant R_{\text{req}}\left(n_{s(y)}\right), \tag{4.30}$$

then the actual reliability $R(G)$ becomes larger than or equal to $R_{\text{req}}(G)$. Considering that the maximum value of $R_{\text{req}}(n_{s(y)})$ is $R_{\max}(n_{s(y)})$ according to (4.29), GMNRA can find an ECU assignment for $n_{s(y)}$ to ensure

$$R(n_{s(y)}) \geqslant R_{\text{req}}(n_{s(y)}) = \frac{R_{\text{req}}(G)}{\prod\limits_{x=1}^{y-1} R(n_{s(x)}) \times \prod\limits_{z=y+1}^{|N|} R_{\text{pre}}(n_{s(z)})}, \tag{4.31}$$

and thereby to ensure

$$\prod\limits_{x=1}^{y-1} R(n_{s(x)}) \times R(n_{s(y)}) \times \prod\limits_{z=y+1}^{|N|} R_{\text{pre}}(n_{s(z)}) = R(G) \geqslant R_{\text{req}}(G). \tag{4.32}$$

4.4.5 GMNRA Algorithm

We present GMNRA algorithm for the parallel application of the embedded system by the above analysis, as shown in Algorithm 6.

GMNRA algorithm preassigns geometric mean-based reliability values to each unassigned task to transfer the reliability requirement of the parallel application to each task. GMNRA then assigns each task to the ECU with the minimum EFT while ensuring its reliability requirement as in Eqs. (4.27)−(4.29). The details of the process are explained as follows.

(1) GMNRA algorithm sorts all tasks in a *task_list* by the descending order of *rank*$_{\text{u}}$ values in line 1.

Algorithm 6 GMNRA Algorithm

Input: $U = \{u_1, u_2, ..., u_{|U|}\}$, G, $R_{\text{req}}(G)$

Output: $R(G)$, and $RT(G)$

1: Sort the tasks in a *task_list* by descending order of $rank_{\text{u}}$ values using Eq. (2.6);
2: Calculate the $R_{\min}(G)$ and $R_{\max}(G)$ using Eqs. (4.10) and (4.9), respectively;
3: Calculate the $GM(G)$ using Eq. (4.25);
4: **for** (z=1; $z \leqslant |N|$; z++) **do**
5:     Calculate $R_{\text{pre}}(n_{s(z)})$ using Eq. (4.26);

6: **endfor**
7: **for** (y=1; $y \leqslant |N|$; y++) **do**
8:     $n_{s(y)} \leftarrow task\_list.out()$;
9:     Calculate $R_{\text{req}}(n_{s(y)})$ using Eqs. (4.27)−(4.29);
10:     **for** (each $u_k \in U$) **do**
11:         Calculate $R(n_{s(y)}, u_k)$ using Eq. (2.1);
12:         **if** ($R(n_{s(y)}, u_k) < R_{\text{req}}(n_{s(y)})$) **then**
13:             **continue**;
14:         **endif**
15:         Calculate $EFT(n_{s(y)}, u_k)$ using Eq. (4.17);

16:     **endfor**
17:     Select $u_{\text{pr}(s(y))}$ with the minimum EFT;
18:     $AFT(n_{s(y)}) \leftarrow EFT(n_{s(y)}, u_{\text{pr}(s(y))})$;

19: **endfor**
20: Calculate $R(G)$ using Eq. (4.6);
21: Calculate $RT(G)$ using Eq. (4.7).

---

(2) GMNRA algorithm calculates the geometric mean of the parallel application by using Eq. (4.25) in lines 2−3.

(3) GMNRA algorithm calculates the preassigned reliability values for the tasks using the Eq. (4.26) in lines 4−6, and all tasks will be traversed in rows 7−19.

(4) GMNRA algorithm calculates the reliability requirement for the current task $n_{s(y)}$ using the Eqs. (4.27)−(4.29) in line 11. Specifically, GMNRA skips ECUs that do not guarantee the reliability requirement of n_i in lines 12−14.

(5) GMNRA algorithm selects the ECU with the minimum EFT for the current task $n_{s(y)}$ under the condition of $R(n_{s(y)}, u_k) \geqslant R_{\text{req}}(n_{s(y)})$ by using Eq. (4.17) in line 15.

(6) GMNRA algorithm calculates the actual reliability and final response time values of the parallel application in lines 20−21, .

The GMNRA algorithm's time complexity is $O(|N|^2 \times |U|)$, which is equal to the time complexity of the HEFT algorithm.

4.4.6 Example of the GMNRA Algorithm

We also use the motivational parallel application in Fig. 2.3 to explain the GMNRA algorithm with the same parameters in Section 4.4.3: $\lambda_1 = 0.0002$, $\lambda_2 = 0.0005$, $\lambda_3 = 0.0009$, and $R_{req}(G) = 0.95$.

Table 4.4: Task assignment generated by GMNRA of the motivational parallel application.

n_i	$R_{req}(n_i)$	$R(n_i,u_1)$	$EFT(n_i,u_1)$	$R(n_i,u_2)$	$EFT(n_i,u_2)$	$R(n_i,u_3)$	$EFT(n_i,u_3)$	$R(n_i)$
n_1	0.997204	**0.997204**	**14**	0.992032	-	0.991933	-	0.997204
n_3	0.993046	**0.997802**	**25**	0.993521	39	0.983045	-	0.997802
n_4	0.987609	0.997403	38	**0.996008**	**31**	0.984816	-	0.996008
n_2	0.983931	**0.997403**	**38**	0.990545	51	0.983931	50	0.997403
n_5	0.991040	0.997603	50	0.993521	44	**0.991040**	**35**	0.991040
n_6	0.992889	**0.997403**	**51**	0.992032	-	0.991933	-	0.997403
n_9	0.984201	0.996406	72	**0.994018**	**66**	0.982161	-	0.994018
n_7	0.990149	**0.998601**	**58**	0.992528	81	0.990149	59	0.998601
n_8	0.987479	**0.999000**	**63**	0.994515	77	0.987479	80	0.999000
n_{10}	0.985703	0.995809	100	**0.996506**	**82**	0.985703	95	0.996506
			$R(G) = 0.965508 > R_{req}(G) = 0.95$, $RT(G) = 82$					

(1) Table 4.4 shows the task assignment generated by GMNRA algorithm of the motivational parallel application. Each row shows the assigned ECU (in boxed) and the reliability value for the task. GMNRA selects the ECU with the minimum EFT for each task without violating its reliability requirement, but their reliability requirement is calculated differently.

(2) By observing Table 4.4, it can be noticed that there is a relatively concentrated trend in reliability values for all tasks after using GMNRA. Specifically, most of the tasks in Table 4.4 have reliability requirements below 0.994, except for n_1. The above concentrated trend can be ascribed to the use of geometric mean. As a result, except for n_1 and n_6, most tasks have enough space to select ECUs with less EFT. Hence, neither high nor low-priority tasks significantly prolong the response time of the parallel application.

(3) Fig. 4.2 shows the task mapping generated by GMNRA of the motivational parallel application, the response time of which is 82.

(4) Table 4.5 compares the actual reliability and final response time when using MRTRR and GMNRA algorithms. Both MRTRR and GMNRA can ensure the reliability requirement of the parallel application, but GMNRA can effectively reduce the response time by a factor of 20 compared to MRTRR. These results show that GMNRA can reduce the imbalance in the assurance of effective reliability requirement and sufficiently reduce the response time. More details between MRTRR and GMNRA are explained in Section 4.6.

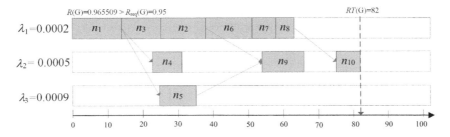

Figure 4.2: Task mapping generated by GMNRA of the motivational parallel application.

Table 4.5: Results of the motivational parallel application using MRTRR and GMNRA.

Algorithm	$R(G)$	$RT(G)$
MRTRR	0.950848	102
GMNRA	0.965508	82

4.5 FAULT TOLERANT RELIABILITY REQUIREMENT VALIDATION

4.5.1 Fault Tolerant Reliability Requirement Assessment

Although GMNRA algorithm can reduce the response time under meeting the reliability requirement of the parallel application, such reliability value cannot be reached if it exceeds $R_{\max}(G)$. This restriction can be interpreted as the existence of only one copy per task, so the maximum achievable reliability value for a parallel application using GMNRA is $R_{\max}(G)$. For example, the maximum achievable reliability value of a motivational parallel application is $R_{\max}(G) = 0.975602$, calculated by Eq. (4.9); if the reliability requirement is 0.976, this requirement cannot be satisfied. Hence, the use of a fault tolerance mechanism is necessary.

In the non-fault tolerant mechanism, the minimum reliability value of the parallel application G is the same as in Eq. (4.10). However, in the fault tolerant mechanism, the maximum reliability value of the parallel application G is changed to

$$R_{\mathrm{fmax}}(G) = \prod_{n_i \in N} R_{\mathrm{fmax}}(n_i), \qquad (4.33)$$

where $R_{\mathrm{fmax}}(n_i)$ is calculated by Eq. (4.3).

4.5.2 Existing Fault Tolerant Functional Safety Validation Algorithms

Taking into account the above-mentioned limitations of non-fault tolerant mechanisms, the HRRM algorithm presented in Reference [62] aims to reduce the redundancy of the parallel application while using fault tolerance to satisfy their reliability requirement. This algorithm is referred to as heuristic replication for response time minimization (HRRTM) in this chapter. Rather than using MRTRR to preassign the maximum reliability value for each unassigned task, HRRTM preassigns the upper reliability value calculated by Eq. (4.23). Then, the reliability requirement of task $n_{s(y)}$ is changed to

$$
R_{\text{req}}(n_{s(y)}) = \frac{R_{\text{req}}(G)}{\prod\limits_{x=1}^{y-1} R(n_{s(x)}) \times \prod\limits_{z=y+1}^{|N|} R_{\text{up}}(n_{s(z)})}
$$

$$
= \frac{R_{\text{req}}(G)}{\prod\limits_{x=1}^{y-1} R(n_{s(x)}) \times \prod\limits_{z=y+1}^{|N|} \sqrt[|N|]{R_{\text{req}}(G)}},
\tag{4.34}
$$

where R_{up} is defined in Eq. (4.23). We use $R_{\text{up}}(n_{s(z)})$ instead of $R_{\text{fmax}}(n_{s(z)})$ because $R_{\text{fmax}}(n_{s(z)})$ is too large and very close to 1 when there are many ECUs in the embedded system, using $R_{\text{fmax}}(n_{s(z)})$ is unbalanced in the fault tolerance mechanism.

4.5.3   Use of Geometric Mean under Fault Tolerance

Even though HRRTM can reduce imbalance, this algorithm can be further optimized. Similar to the non-fault tolerant algorithm, we use the geometric mean shown in Eq. (4.35) to further reduce the imbalance.

$$
GM(G) = \sqrt[|N|]{\frac{R_{\text{req}}(G)}{R_{\min}(G) \times R_{\text{fmax}}(G)}}.
\tag{4.35}
$$

Given that $R_{\text{fmax}}(G)$ is close to 1 and is imbalanced, we replace this parameter with $R_{\text{up}}(G)$. This replacement refers to the HRRTM where $R_{\text{fmax}}(n_{s(z)})$ is replaced with $R_{\text{up}}(n_{s(z)}) = \sqrt[|N|]{R_{\text{req}}(G)}$. The geometric mean of the parallel application is changed to

$$
GM'(G) = \sqrt[|N|]{\frac{R_{\text{req}}(G)}{R_{\min}(G) \times R_{\text{up}}(G)}}.
\tag{4.36}
$$

Similar to the non-fault tolerant algorithm, the preassigned reliability value of $n_{s(z)}$ is calculated by

$$
R_{\text{pre}}(n_{s(z)}) = GM'(G) \times R_{\min}(n_{s(z)}) \times R_{\text{fmax}}(n_{s(z)}).
\tag{4.37}
$$

As $R_{\text{fmax}}(n_{s(z)})$ is replaced with $R_{\text{up}}(n_{s(z)})$ in HRRTM, we also let $R_{\text{fmax}}(n_{s(z)})$ be replaced with $R_{\text{up}}(n_{s(z)})$ in this subsection, and the preassigned reliability value of $n_{s(z)}$ is changed to

$$
\begin{aligned}
R'_{\text{pre}}(n_{s(z)}) &= GM'(G) \times R_{\min}(n_{s(z)}) \times R_{\text{up}}(n_{s(z)}) \\
&= \sqrt[|N|]{\frac{R_{\text{req}}(G)}{R_{\min}(G) \times R_{\text{up}}(G)}} \times R_{\min}(n_{s(z)}) \times R_{\text{up}}(n_{s(z)}) \\
&= \sqrt[|N|]{\frac{R_{\text{req}}(G)}{R_{\min}(G)}} \times R_{\min}(n_{s(z)}).
\end{aligned}
\tag{4.38}
$$

Hence, we also place the preassigned reliability values of the unassigned tasks in the fault tolerant mechanism in a concentrated trend. According to the above process, the reliability requirement in assigning $n_{s(y)}$ is calculated by

$$R'_{\text{req}}(n_{s(y)}) = \frac{R_{\text{req}}(G)}{\prod\limits_{x=1}^{y-1} R(n_{s(x)}) \times \prod\limits_{z=y+1}^{|N|} R'_{\text{pre}}(n_{s(z)})}. \qquad (4.39)$$

Given that the minimum reliability of $n_{s(y)}$ is $R_{\min}(n_{s(y)})$, $R'_{\text{req}}(n_{s(y)})$ must be updated to

$$R'_{\text{req}}(n_{s(y)}) = \max\left\{ R_{\min}(n_{s(y)}), R'_{\text{req}}(n_{s(y)}) \right\}. \qquad (4.40)$$

Given that the upper reliability of $n_{s(y)}$ is $R_{\text{up}}(n_{s(y)})$, $R'_{\text{req}}(n_{s(y)})$ must be further updated to

$$R'_{\text{req}}(n_{s(y)}) = \min\left\{ R_{\text{up}}(n_{s(y)}), R'_{\text{req}}(n_{s(y)}) \right\}. \qquad (4.41)$$

For any task $n_{s(y)}$, if and only if

$$R\left(n_{s(y)}\right) \geqslant R'_{\text{req}}\left(n_{s(y)}\right), \qquad (4.42)$$

then the actual reliability $R(G)$ becomes larger than or equal to $R'_{\text{req}}(G)$. Considering that the maximum value of $R'_{\text{req}}(n_{s(y)})$ is $R_{\text{up}}(n_{s(y)})$ according to Eq. (4.41), and $R_{\text{up}}(n_{s(y)})$ is the actual maximum reliability value of each task according to Eq. (4.23), GMFRA can find an ECU assignment for $n_{s(y)}$ to satisfy

$$R(n_{s(y)}) \geqslant R'_{\text{req}}(n_{s(y)}) = \frac{R_{\text{req}}(G)}{\prod\limits_{x=1}^{y-1} R(n_{s(x)}) \times \prod\limits_{z=y+1}^{|N|} R'_{\text{pre}}(n_{s(z)})}, \qquad (4.43)$$

thereby ensuring

$$\prod\limits_{x=1}^{y-1} R(n_{s(x)}) \times R(n_{s(y)}) \times \prod\limits_{z=y+1}^{|N|} R'_{\text{pre}}(n_{s(z)}) = R(G) \geqslant R_{\text{req}}(G). \qquad (4.44)$$

4.5.4 Optimizing Response Time

After determining the reliability requirement for each task, we need to select approximate copies for each task to satisfy its reliability requirement and reduce its finish time.

Two main types of primary-backup replications exist: passive replication and active replication [62, 75, 76]. For passive replication, tasks are rescheduled whenever ECUs are unable to continue on the same or backup ECUs. For active replication, each task is replicated on multiple ECUs at the same time, and the task succeeds if at least one copy is completed.

Two types of fault tolerant schedules can be used, namely, compile-time strict schedule and run-time general schedule [62]. For compile-time strict scheduling, each task must wait for all copies of its predecessors to complete (both successes and failures) before it can start execution. For runtime general scheduling, the execution of each task can start as soon as one copy of each predecessor completes successfully. Similar to References [62, 75, 76], this chapter uses active replication and compile-time strict scheduling in the design phase to obtain predictable scheduling results.

Based on active replication and strict scheduling, the EFT of n_i to u_k in the fault tolerant mechanism is calculated by

$$EFT(n_i^\beta, u_k) = EST(n_i^\beta, u_k) + w_{i,k}, \tag{4.45}$$

where n_i^β denotes the β-th selected copy of n_i. The EST of n_i on u_k is updated to

$$\begin{cases} EST(n_{\text{entry}}, u_k) = 0, \\ EST(n_i, u_k) = \max \left\{ \begin{aligned} & avail[k], \\ & \max_{n_h \in pred(n_i), \alpha \in [1, num_h]} \left\{ AFT(n_h^\alpha) + c'_{h,i} \right\} \end{aligned} \right\}, \end{cases} \tag{4.46}$$

where num_h denotes the assigned number of copies of n_h, while num_i represents the actual number of assigned copies of n_i. Hence, the reliability of n_i is updated to

$$R(n_i) = 1 - \prod_{\beta=1}^{num_i} \left(1 - R\left(n_i^\beta, u_{\text{pr}(n_i^\beta)} \right) \right), \tag{4.47}$$

where $u_{\text{pr}(n_i^\beta)}$ represents the assigned ECU of n_i^β.

We iteratively assign the copies of current task n_i to the available ECUs with minimum EFTs until the reliability requirement of n_i is satisfied. That is, the assigned ECU $u_{\text{pr}(n_i^\beta)}$ and corresponding $EFT(n_i, u_{\text{pr}(n_i^\beta)})$ for n_i in each iteration is determined as

$$AFT(n_i) = EFT(n_i, u_{\text{pr}(n_i^\beta)}) = \min_{u_k \in U, u_k \text{ is available}} \left\{ EFT(n_i, u_k) \right\}, \tag{4.48}$$

where "u_k is available" means that no other copies of n_i have been assigned to u_k because a replication on the same ECU is not allowed according to the active replication [62, 75, 76].

The final response time of the parallel application is considered the AFT of the copy of the exit task n_{exit}; this copy has the maximum AFT among all copies of n_{exit}. Hence, we have

$$RT(G) = AFT(n_{\text{exit}}) = \max_{\beta \in [1, num_{\text{exit}}]} \left\{ AFT(n_{\text{exit}}^\beta) \right\}. \tag{4.49}$$

4.5.5 GMFRA Algorithm

Based on the above analysis, we design the GMFRA algorithm for the parallel application of embedded systems, as shown in Algorithm 7.

GMFRA algorithm also preassigns reliability values based on the geometric mean to each unassigned task to transfer the reliability requirement of the parallel application to each task. In contrast to GMNRA, GMFRA uses the upper reliability $R_{\text{up}}(n_{s(z)})$ instead of the maximum reliability $R_{\text{fmax}}(n_{s(z)})$ to reduce the imbalance

---

Algorithm 7 GMFRA Algorithm

Input: $U = \{u_1, u_2, ..., u_{|U|}\}$, G, $R_{\text{req}}(G)$
Output: $R(G)$, and $RT(G)$

1: Sort the tasks in a *task_list* by descending order of $rank_u$ values using Eq. (2.6);
2: Calculate the $R_{\min}(G)$ using Eq. (4.10);
3: Calculate the $GM(G)'$ using Eq. (4.36);
4: **for** $(z=1; z \leqslant |N|; z++)$ **do**
5:      Calculate $R'_{\text{pre}}(n_{s(z)})$ using Eq. (4.38);
6: **endfor**
7: **for** $(y=1; y \leqslant |N|; y++)$ **do**
8:      $n_{s(y)} \leftarrow task\_list.out()$;
9:      Calculate $R'_{\text{req}}(n_{s(y)})$ using Eqs. (4.39)$-$(4.41);
10:      **for** (each ECU $u_k \in U$) **do**
11:          Calculate $R(n_{s(y)}, u_k)$ using Eq. (2.1);
12:          Calculate $EFT(n_{s(y)}, u_k)$ using Eq. (4.45);
13:      **endfor**
14:      $R(n_{s(y)}) \leftarrow 0$;
15:      **while** $(R(n_{s(y)}) < R'_{\text{req}}(n_{s(y)}))$ **do**
16:          Assign $n_{s(y)}$'s copy to the available and ECU with the minimum EFT by Eq. (4.48);
17:          Calculate $R(n_{s(y)})$ using Eq. (4.47);
18:      **endwhile**
19: **endfor**
20: Calculate the reliability $R(G)$ using Eq. (4.6);
21: Calculate $RT(G)$ using Eq. (4.49).

---

in the fault tolerance mechanism. Then, GMFRA iteratively assigns a copy of the current task $n_{s(y)}$ to the available ECU with the minimum EFT until the reliability requirement of $n_{s(y)}$ is satisfied.

(1) GMFRA algorithm sorts all tasks in a list *task_list* by the descending order of $rank_u$ values in line 1.

(2) GMFRA algorithm calculates the geometric mean of the parallel application by using Eq. (4.36) in lines 2−3.

(3) GMFRA algorithm calculates the preassigned reliability value of each task by using Eq. (4.38) in lines 4−6. All tasks will be traversed in lines 7−19.

(4) GMFRA algorithm calculates the reliability requirement of the current task $n_{s(y)}$ by using Eqs. (4.39)−(4.41) in line 9.

(5) GMFRA algorithm iteratively assigns the copies of current task $n_{s(y)}$ to the available ECUs with the minimum EFTs until its reliability requirement is satisfied in lines 15−18.

(6) GMFRA algorithm calculates the actual reliability and response time of the parallel application in lines 20 and 21, respectively.

The GMFRA algorithm's time complexity is also $O(|N|^2 \times |U|^2)$ because it increases one $O(|U|)$ in calculating EFT than GMNRA algorithm.

4.5.6 Example of the GMFRA Algorithm

The maximum reliability value for the motivational parallel applications under a non-fault tolerant mechanism is 0.975602, and this example assumes that the reliability requirement is not less than 0.988. From the previous sections, it is clear that the use of both MRTRR and GMNRA algorithms cannot satisfy the reliability requirement, while GMFRA algorithm can achieve the goal. For three ECUs, the failure ratios are still $\lambda_1 = 0.0002$, $\lambda_2 = 0.0005$, and $\lambda_3 = 0.0009$. Table 4.4 shows the task assignment of the motivational parallel application using the GMFRA algorithm.

Table 4.6: Task assignment generated by GMFRA of the motivational parallel application.

n_i	$R_{req}(n_i)$	$R(n_i, u_1)$	$EFT(n_i, u_1)$	$R(n_i, u_2)$	$EFT(n_i, u_2)$	$R(n_i, u_3)$	$EFT(n_i, u_3)$	$R(n_i)$
n_1	0.998793	**0.997204**	**14**	0.992032	16	**0.991933**	**9**	0.999977
n_3	0.998150	**0.997802**	**32**	**0.993521**	**39**	0.983045	45	0.999986
n_4	0.994540	**0.997403**	**45**	0.996008	47	**0.984816**	**40**	0.999961
n_2	0.990077	**0.997403**	**58**	0.990545	58	0.983931	58	0.997403
n_5	0.995301	0.997603	70	**0.993521**	**52**	**0.991040**	**50**	0.999942
n_6	0.998793	0.997403	71	**0.992032**	**68**	**0.991933**	**59**	0.997403
n_9	0.992665	**0.996406**	**83**	0.994018	86	0.982161	94	0.996406
n_7	0.998002	0.998601	90	**0.992528**	**83**	**0.990149**	**73**	0.999926
n_8	0.997141	**0.999000**	**88**	0.994515	94	0.987479	97	0.999000
n_{10}	0.995410	0.995809	121	**0.996506**	**106**	0.985703	116	0.996506
			$R(G) = 0.989087 > R_{req}(G) = 0.988$, $RT(G) = 106$					

(1) Unlike the MRTRR and GMNRA algorithms (non-fault tolerant), which select only one copy, the GMFRA algorithm (fault tolerant) selects one or more copies for each task in each row.

(2) GMFRA first considers n_1 according to the determined task priority. The reliability requirement of n_1 is 0.998793 calculated by Eqs. (4.39)−(4.41). The reliability values of n_1 on u_1, u_2, and u_3 are 0.997204, 0.992032, and 0.991933 calculated by Eq. (2.1). The EFT values of n_1 on u_1, u_2, and u_3 are 14, 16, and 9 calculated by Eq. (4.17). GMFRA first selects u_3 for n_1 because its minimum EFT is 9, and its reliability value is 0.991933. However, 0.991933 cannot meet the reliability requirement of 0.998793. Hence, GMFRA should further select the available u_1 for n_1 with the second minimum EFT of 14. Accordingly, reliability value of n_1 reaches 0.999977 calculated by Eq. (4.5) ($0.999977 = 1 - (1 - 0.991933) \times (1 - 0.997204)$). Correspondingly, n_1's task mapping on u_3 and u_1 is shown in Fig. 4.3. In summary, GMFRA iteratively

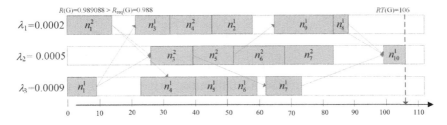

Figure 4.3: Task mapping generated by GMFRA of the motivational parallel application.

selects available copies and ECUs with minimum EFTs for each task until its reliability requirement is satisfied.

(3) The actual response time and reliability value of the motivational parallel application is 106 and 0.989087, which are obtained by Eqs. (4.6) and (4.7), respectively. This example indicates that GMFRA can satisfied higher reliability requirement through the use of fault tolerance. The final task assignment generated by GMFRA for the motivated parallel application G is shown in Fig. 4.3.

4.6 EXPERIMENTS FOR FUNCTIONAL SAFETY VALIDATION ALGORITHMS GMNRA AND GMFRA

To validate the effectiveness of the presented GMNRA and GMFRA algorithms, this section uses the MRTRR [61] and HRRTM [62] algorithms as comparisons.

The performance metrics to be compared in this experiment should be the actual reliability values and response times of the parallel application. Given that the reliability requirement can be met within the individual evaluations, no specific reliability values are provided in this chapter.

We also use parameter values from a real parallel application of an automotive embedded system as experimental data. Each task's failure rate is in the range of $10^{-6} - 9 \times 10^{-6}$ in time unit of $1\,\mu s$, and the messages of the task's WCETs and WCRTs fall in the range of $100-400\,\mu s$. The above values are generated from a uniform distribution.

Our target is to compute the assigned ECU, and task mapping) for each task using the presented algorithm. The following implementation phase allows for parallel applications based on the established task mapping. Hence, the parallel applications in this section will be tested on a simulated system according to the above application parameter values to indicate a real deployment.

4.6.1 Real-Life Parallel Application

We take the real-life automotive parallel application shown in Fig. 2.9 from Reference [61]. This application consists of six functional blocks, namely, engine controller with seven tasks $(n_1 - n_7)$, automatic gear box with four tasks $(n_8 - n_{11})$, anti-locking brake system with six tasks $(n_{12} - n_{17})$, wheel angle sensor with two tasks $(n_{18} - n_{19})$, suspension controller with five tasks $(n_{20} - n_{24})$, and body work with seven tasks $(n_{25} - n_{31})$.

Experiment 1. The purpose of this experiment is to compare the response time values of real-life parallel applications. 1000 automotive parallel applications are generated according to the above parameter values. For highly safety-critical automotive parallel applications with an exposure level of E1, a reliability of 0.999999 [4] is required. However, according to our calculations, for all 1000 applications, the maximum reliability value without using fault tolerance is between 0.992 and 0.993. Therefore, we set the reliability requirement change from 0.980 to 0.992 in increments of 0.002. According to Table 3.9, the reliability requirements of 0.90 and 0.92 fall in the range of exposure levels E2 and E1, respectively. The results of experiment are shown in Fig. 4.4

(1) In 1000 comparisons, the number of times the GMNRA obtained a faster response time than the MRTRR was 711-896. That is, the GMNRA algorithm has a 70%–90% probability of achieving faster response times than MRTRR while still meeting reliability requirement.

(2) In all cases, the average response time generated by GMNRA are faster than those generated by MRRTR. Using both MRRTR and GMNRA algorithms, the average response time increases with increasing reliability requirement. In summary, the average response time of GMNRA can be reduced by 13.02%−36.90% compared to MRTRR.

(3) We can observe that the number of times when MRTRR and GMNRA times are shorter does not add up to 1000. Because MRTRR and GMNRA algorithms obtain equal response time values in a few cases.

Experiment 2. Considering the reliability requirement of 0.999999 or even higher, non-fault tolerant MRTRR and GMNRA do not work in this case. Hence, we use the fault tolerant HRRTM and GMFRA algorithms to analyze the results. We set the reliability requirement change from 0.999 to 0.99999999. All these reliability requirement values fall within the range of exposure level E1 in Table 3.9. The results of the experiment are shown in Fig. 4.5.

(1) In 1000 comparisons, the probability of GMFRA obtaining a shorter response time than MRTRR on all scales is 431−843, while the probability of HRRTM obtaining a shorter response time than GMFRA is 36−337. The advantage of GMFRA algorithm is evident when the reliability requirement is greater than or equal to 0.999999. For instance, when the reliability requirement is 0.99999999, the probability of GMFRA algorithm obtaining a faster response time than HRRTM is 84.3%.

(2) In all cases, the average response times calculated by using GMFRA are still faster than those calculated by using HRRTM. The average response times

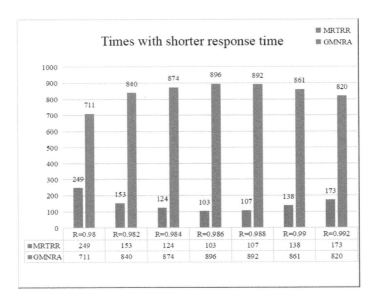

(a) Times with shorter response time.

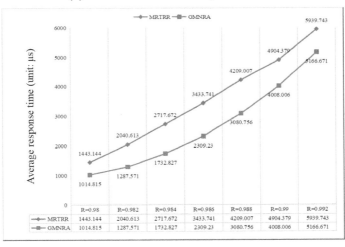

(b) Average response time(unit: μ s).

Figure 4.4: Results generated by MRTRR and GMNRA of 1000 real-life parallel applications.

increase with the reliability requirement of using HRRTM and GMFRA algorithms. In summary, GMFRA algorithm can reduce response times by as much as 9.89% compared to HRRTM algorithm. Even though this percentage is not very high, the result is impressive for highly safety applications, as automotive parallel applications are time accurate to the microsecond. In other words, automotive parallel applications are time-sensitive. For instance, when the

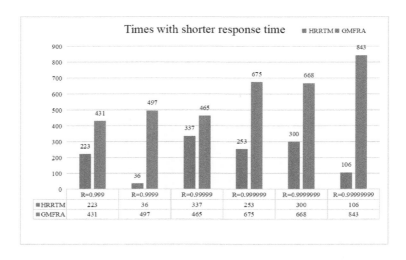

(a) Times with shorter response time.

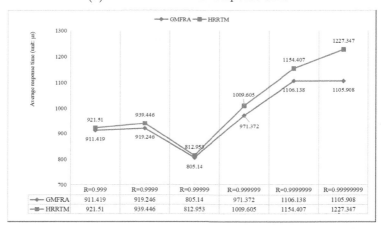

(b) Average response time(unit: μs).

Figure 4.5: Results generated by HRRTM and GMFRA of 1000 real-life parallel applications.

reliability requirement is 0.99999999, the reduced average response time is 121.439 μs, which already reflects a good improvement.

(3) Fig. 4.5 shows that the number of times with shorter time for HRRTM and GMFRA algorithms do not add up to 1000 because HRRTM and GMFRA algorithms also obtain equal response times in a few cases.

4.6.2 Synthetic Parallel Application

Experiment 3. Using a single real-life parallel application of an automotive embedded system to confirm the effectiveness of the proposed algorithms is considered insufficient. Therefore, we use additional synthetic parallel applications with the same parameter values of real-life parallel applications to analyze the experiment results. We can obtain randomly generated parallel applications by using a task graph generator [1]. The task graph parameters are as follows:

(1) The communication rate set is $\{0.1, 0.5, 1.0, 2.0, 5.0\}$.

(2) The heterogeneous factor set of ECUs is $\{0.2, 0.4, 0.5, 0.6, 0.8, 1.0\}$. The larger the factor, the higher the heterogeneity.

(3) The shape parameter is a random number that takes values in the range $[0.35, \sqrt{|N|/3}]$. We set the shape parameter be set to $\{0.35, 1.0, 2.0, 3.0, 4.08, 4.47, 4.83, 5.16, 5.47, 5.77\}$. Note that the maximum shape factors are 4.08 and 5.77 for 50 and 100 tasks, respectively.

The parameters for each parallel application are chosen randomly within the above parameters for synthetic parallel applications. We let the reliability requirement vary from 0.999 to 0.99999999 to observe the results using HRRTM and GMFRA algorithms. The results of experiment are shown in Fig. 4.6.

Fig. 4.6(a) shows that GMFRA algorithm has a significant advantage over HRRTM algorithm in all cases. GMFRA algorithm has a $645-818$ probability of obtaining a shorter response time, while HRRTM has a $282-353$ probability. GMFRA can reduce the average response time by as much as 160 μs, which is a very good improvement for time sensitive automotive parallel applications. In summary, all of the above experiments reflect that GMNRA is more effective than MRTRR in ensuring high reliability requirement.

4.7 CONCLUDING REMARKS

This chapter introduces GMNRA and GMFRA algorithms for a parallel application of an automotive embedded system in the design phase, with a focus on automotive functional safety validation. We introduce geometric mean on which the preassigned reliability values are implemented. Regarding the unbalanced reliability values preassigned to unassigned tasks in MRCRG, we propose the GMNRA algorithm in which the geometric mean-based reliability values are preassigned to unallocated tasks. Regarding the unbalanced reliability values preassigned to unallocated tasks in HRRM, we propose the GMFRA algorithm, in which the geometric mean-based reliability values are preassigned to unallocated tasks. These two algorithms are applicable to non-fault tolerant and fault tolerant mechanisms, respectively.

Compared to individual state of the art methods, GMNRA and GMFRA can effectively reduce the response time of a parallel application of an automotive embedded system while satisfying its reliability requirement. In a broad sense, GMNRA and GMFRA can effectively enhance automotive functional safety during the design phase.

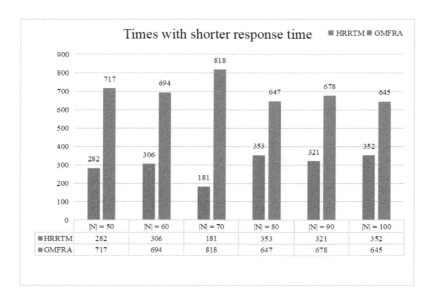

(a) Times with shorter response time.

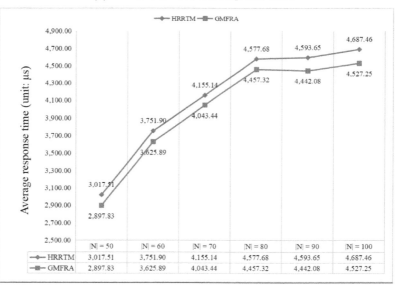

(b) Average response time(unit: μs).

Figure 4.6: Results generated by HRRTM and GMFRA of 1000 synthetic applications with reliability requirement of 0.99.

Compared to other state-of-the-art methods, GMNRA and GMFRA algorithms effectively reduce the response time of a parallel application of an automotive embedded system while not violating reliability requirements. In other words, GMNRA and GMFRA algorithms can effectively ensure automotive functional safety in the design phase.

II

SAFETY-AWARE COST OPTIMIZATION

II

CHAPTER 5

Hardware Cost Optimization

I NDUSTRIAL embedded system is usually cost-sensitive, and reducing system hardware costs can lead to higher profits. The functional safety requirements must be satisfied according to functional safety standards. In this chapter, we propose two hardware cost optimization approaches. The first approach presents the progressive hardware cost optimization (PHCO), enhanced PHCO (EPHCO), and simplified EPHCO (SEPHCO) algorithms step by step for a parallel application while satisfying the functional safety requirements. The second approach presents the cost-effectiveness-driven hardware cost optimization algorithm (CEHCO) for a parallel application while meeting the functional safety requirements. CEHCO combines CEHCO1 and CEHCO2 algorithms proposed in this chapter, thus achieving both robust cost optimization capabilities and superior time efficiency. Finally, we perform hardware cost optimization experiments with real-life and synthetic parallel applications for automotive embedded systems to verify the performance and efficiency of PHCO, EPHCO, SEPHCO, and CEHCO algorithms.

5.1 INTRODUCTION

5.1.1 Progressive Hardware Cost Optimization

Reducing the hardware cost can result in significant savings in industrial production costs and obtain high profits. While distributed architecture greatly reduce the hardware cost of the wiring harness, the number of ECUs required for application execution still incurs significant hardware costs. In addition, as the size of the automotive applications continues to grow, the cost of the required ECU hardware increases accordingly. At present, the unit price of ECUs for CAN interfaces is from $5 to $100 [27]. Actually, some of the unnecessary ECUs can be removed, as long as such operations do not interfere with the correct operation of the automotive application. Hence, hardware costs can be reduced by minimizing the number of ECUs in automotive embedded systems [22].

A safety-critical parallel application of embedded systems must meet related functional safety standards. Exceptional executions and run timeouts are considered to be common random hardware failures and system failures, respectively. In other words, the normal reliability requirement and real-time requirement must be met to ensure

DOI: 10.1201/9781003391517-5

the functional safety of a parallel application of an embedded system according to the relevant functional safety standards. Therefore, hardware costs need to be reduced while meeting functional safety requirements, including real-time and reliability requirements.

In this chapter, our main goal is to optimize hardware costs while meeting the functional safety requirements of the parallel application of an embedded system. The first group of techniques in this chapter is summarized as follows:

(1) We propose a progressive hardware cost optimization (PHCO) algorithm by iteratively removing the ECUs, and it can yield minimal hardware cost while ensuring functional safety requirements of the parallel application of an embedded system.

(2) To address the problem that some of the results generated by PHCO can satisfy the real-time requirement but may not satisfy the reliability requirement, we propose a reliability enhancement (RE) algorithm. The RE algorithm increases the reliability values of parallel applications without violating the inter-task precedence constraint and real-time requirement. Then we propose an enhanced PHCO (EPHCO) algorithm based on the RE algorithm to improve the possibility of satisfying the functional safety requirements.

(3) The drawback of PHCO and EPHCO is their high time complexity, while large-scale parallel applications require a large amount of computational work, so we propose a simplified EPHCO (SEPHCO) algorithm to accommodate hardware cost design optimization for large-scale parallel applications.

5.1.2 Cost-Effectiveness-Driven Hardware Cost Optimization

In the IEC 61508 standard, the probability of a hazardous failure per hour must be larger than or equal to 10^{-9} and less than 10^{-8} in safety integrity level (SIL) 4 [26]. As long as one of the real-time and reliability requirement is not satisfied, the corresponding personal safety is not guaranteed.

EPHCO iteratively closes some ECUs from all open ECUs until the functional safety requirements of the parallel application cannot be met (i.e., the opened-to-closed approach). SEPHCO simplifies the iteration details to overcome EPHCO's time inefficiencies. EPHCO algorithm is highly cost-optimized but less time efficient. The performance of EPHCO and SEPHCO is unacceptable in a large-scale heterogeneous embedded system (e.g., smart grid) configured with a large number of ECUs (at least hundreds of ECUs). The size of parallel applications executing in such systems is relatively large (at least thousands of tasks). Massively parallel applications of embedded systems require a large number of ECUs involved in large-scale computations. However, using EPHCO and SEPHCO can lead to time inefficiencies and thus long lifecycles. To address these issues, we need to propose a new approach that can implement both strong cost optimization capabilities and superior time efficiency in a large-scale heterogeneous embedded system.

We devise the concept of cost-effectiveness and propose a cost-effectiveness-driven hardware cost optimization (CEHCO) algorithm. CEHCO addresses the problem of simultaneous compatibility with robust cost optimization and superior time efficiency in a large-scale heterogeneous embedded system through cost-performance-driven closed-to-opened and opened-to-closed approaches.

(1) We propose the first cost-effectiveness-driven hardware cost optimization (CE-HCO1) algorithm. CEHCO1 repeatedly selects the most cost-effective ECU to open until the functional safety requirements are met (i.e., closed-to-opened).

(2) We design a second cost-effectiveness-driven hardware cost optimization (CE-HCO2) algorithm. CEHCO2 repeatedly selects the ECU with the lowest cost-effectiveness ratio to shut down until functional safety requirements cannot be met. It also further optimizes the hardware cost on top of CEHCO1 without losing time efficiency.

(3) We propose the CEHCO algorithm by combining the CEHCO1 and CEHCO2 algorithms. Given that CEHCO2 can be engaged several times until a stable value is reached. By significantly reducing the number of iterations, CEHCO solves the problem of time inefficiency of opened-to-closed methods. By using the UFFSV proposed in Section 2.5.4, CEHCO achieves a powerful cost optimization capability.

5.2 RELATED WORK

We mainly review related research on functional safety and cost optimization of a parallel application of embedded systems.

Reliability maximization and response time minimization are contradictory, and optimizing them is a dual-objective minimization problem [20, 62]. Reference [20] proposed a bi-objective scheduling heuristic (BSH) to obtain an approximate Pareto curve of non-dominated solutions. In above mentioned Pareto curve, designers can verify the functional safety requirements by searching the point set that meets the real-time and reliability requirements simultaneously. Reference [62] studied the same problem as RR and MaxRe, and it presented the heuristic replication for redundancy minimization (HRRM) method. HRRM has shown significant improvements in reducing resource costs. Due to the limited resource of the embedded system, fault-tolerance may be unsuitable [61]. Reference [61] presented resource cost optimization to meet the reliability requirement by transferring the reliability requirement of a parallel application of the embedded system to each task without using fault-tolerance.

In addition to the resource cost, the development cost and hardware cost design optimization for the safety-critical parallel application is studied in References [19, 22, 56]. References [19, 56] proposed minimizing development cost to meet the real-time requirement for the parallel application of the embedded system. Reference [22] devised hardware cost minimization to meet the real-time and safety requirements for a parallel application by proposing integer linear programming and heuristics, respectively. Although hardware cost is introduced in Reference [22], it focuses on safety requirement rather than reliability requirement.

5.3 MODELS AND PROBLEM STATEMENT

We use the same embedded system architecture, parallel application model, and reliability model as in Section 2.3.1. A parallel application is represented by a DAG $G=(N, W, M, C)$. Tables 5.1 and 5.2 list the abbreviations and notations that are used in this chapter.

Table 5.1: Abbreviations in this chapter.

Abbreviation	Definition
FFT	Fast Fourier Transform
GE	Gaussian Elimination
HC	Hardware Cost
SIL	Safety Integrity Level
HCO	Hardware Cost Optimization
IHCO	Iterated Hardware Cost Optimization
PHCO	Progressive Hardware Cost Optimization
EPHCO	Enhanced PHCO
SEPHCO	Simplified EPHCO
CEHCO	Cost-Effectiveness-driven Hardware Cost Optimization

Table 5.2: Notations in this chapter.

Notation	Definition
$price_k$	Unit price the ECU u_k
$HC(G)$	Hardware cost of the parallel application G
$CE(u_k)$	Cost-Effectiveness of u_k
U_{opened}	Opened ECU set in parallel applications
U_{closed}	Closed ECU set in parallel applications

5.3.1 Hardware Cost Model

All ECUs have individual unit prices. Therefore, let $\{price_1, price_2, ..., price_{|U|}\}$ represent the set of unit prices of ECUs. Table 5.3 shows that the example of the hardware costs of the ECUs u_1, u_2, u_3 are 20, 10, and 30, respectively.

The parallel application's hardware cost is the sum of those of all open ECUs, and it is calculated by

$$HC(G) = \sum_{u_k \in U} price_k. \tag{5.1}$$

Table 5.3: ECU parameters for the motivational parallel application.

ECUs/Parameters	$price_k$	λ_k
u_1	20	0.00055
u_2	20	0.0005
u_3	30	0.0004

5.3.2 Problem Statement

The main problem addressed in this chapter is to search the ECU assignment schemes for all tasks in order to minimize the hardware cost:

$$HC(G) = \sum_{u_k \in U_{\text{active}}} price_k, \tag{5.2}$$

while ensuring that real-time requirement is satisfied:

$$RT(G) = AFT(n_{\text{exit}}) \leqslant RT_{\text{req}}(G), \tag{5.3}$$

and actual reliability is greater than reliability requirement:

$$R(G) \geqslant R_{\text{req}}(G). \tag{5.4}$$

$RT_{\text{req}}(G)$ represents the real-time requirement, and $R_{\text{req}}(G)$ represents reliability requirement. $AFT(n_{\text{exit}})$ represents the actual finish time of n_{exit} of the parallel application G. Hardware cost optimization under functional safety requirements in this chapter is an NP-hard optimization problem, since scheduling tasks with optimal quality-of-service requirements in multiple ECUs is an NP-hard problem [59].

5.4 PROGRESSIVE HARDWARE COST OPTIMIZATION

5.4.1 IHCO Algorithm

Since our goal is to minimize the hardware cost of the parallel application of an embedded system, reducing the number of ECUs is an approach worth considering. Hence, we should reduce the number of ECUs as much as possible (i.e., remove some of ECUs) and migrate the tasks of the parallel application to the remaining ECUs, while meeting the functional safety requirements.

Some studies use the method of removing ECUs to reduce energy consumption. The method proposed in Reference [57] is iteratively removing as many ECUs as possible with a small number of tasks assigned to reduce energy consumption until real-time requirement cannot be met. The main idea of Reference [57] can also be applied to hardware cost optimization, where the reduction of the number of ECUs is achieved by removing as many ECUs with high hardware costs as possible, but only if the functional safety requirements of the parallel application are assured. We call this approach iterative hardware cost optimization (IHCO). In the motivational parallel application, ECU u_3 has the highest cost (i.e., 30) among $\{u_1, u_2, u_3\}$, as shown in Fig. 5.1, we must remove u_3 from $\{u_1, u_2, u_3\}$ and allocate all the tasks of the motivational parallel application to u_1 or u_2. Of course, we can set u_3 to sleep instead of actually removing it from the system during the design phase.

Fig. 5.1 shows the task mapping generated by HEFT algorithms on u_1 and u_2 when u_3 is in the sleep state. The response time is 82, which can meet the deadline $(D(G) = 100)$. However, the reliability value is 0.946816, which cannot meet the reliability requirement of 0.95. That is, removing u_3 from the ECU set is infeasible in this case. IHCO cannot get a valid hardware cost for the parallel application. Hence, iteratively removing as many ECUs with a high hardware cost is not an optimized design and we need to present better methods.

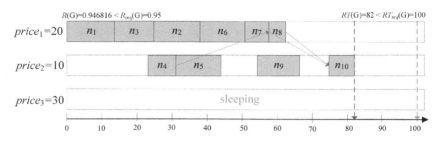

Figure 5.1: Task mapping-based HEFT on u_1 and u_2 when u_3 is in a sleep state.

5.4.2 PHCO Algorithm

In order to improve the limitations of IHCO, we present a progressive hardware cost optimization method. Similar to Fig. 5.1, if we set u_1 or u_2 in the sleep state, the task mappings are shown in Figs. 5.2 and 5.3, respectively.

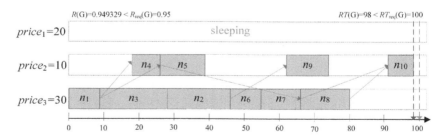

Figure 5.2: Task mapping-based HEFT on u_2 and u_3 when u_1 is in the sleep state.

Compared to Fig. 5.1, we find that higher reliability values are obtained in Figs. 5.2 and 5.3 under real-time requirement. However, Figs. 5.2 and 5.3 still do not meet the reliability requirement of 0.95, and their results suggest that removing the ECU using the HEFT algorithm does not necessarily increase the reliability value of the parallel application. Based on the above analysis, we propose the progressive approach called PHCO as shown in Algorithm 8.

EHCO minimizes hardware costs by iteratively removing ECUs from active ECUs U_{active} while meeting functional safety requirements of the parallel application of an embedded system. The details of Algorithm 8 are explained as follows.

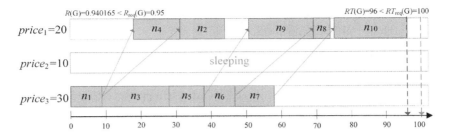

Figure 5.3: Task mapping-based HEFT on u_1 and u_3 when u_2 is in the sleep state.

Algorithm 8 PHCO Algorithm

Input: $U = \{u_1, u_2, ..., u_{|U|}\}$, G, $RT_{\text{req}}(G)$, and $R_{\text{req}}(G)$
Output: $R(G)$, $RT(G)$, and $HC(G)$

1: $U_{\text{active}} \leftarrow P$;
2: Call the HEFT on U_{active} to obtain the initial $RT(G) \leftarrow LB(G)$, $HC(G)$, and $R(G)$;
3: **if** $R(G) < R_{\text{req}}(G))$ or $(RT(G) > RT_{\text{req}}(G)$ **then**
4:     **return** false;
5: **endif**
6: **while** (U_{active} is not NULL) **do**
7:     **for** (each active ECU u_k in U_{active}) **do**
8:         Set the u_k state to sleep;
9:         Invoke the HEFT on $U_{\text{active}} - \{u_k\}$ to calculate $R_k(G)$, $RT_k(G)$, and $HC_k(G)$;
10:     **end for**
11:     **if** (no result satisfies $RT_k(G) \leqslant RT_{\text{req}}(G)$ and $R_k(G) \geqslant R_{\text{req}}(G)$) **then**
12:         **return** true;
13:     **else**
14:         Calculate the ECU u_{min} that has $HC_{\text{min}}(G)$ using Eq. (5.5);
15:         $U_{\text{active}} \leftarrow (U_{\text{active}} - \{u_{\text{min}}\})$;
16:         $R(G) \leftarrow R_{\text{min}}(G)$, $RT(G) \leftarrow RT_{\text{min}}(G)$, and $HC(G) \leftarrow HC_{\text{min}}(G)$;
17:     **endif**
18: **endwhile**

---

(1) Let all the ECUs' states be active in line 1.

(2) PHCO invokes the HEFT algorithm on all the ECUs to get the $R_k(G)$, $RT_k(G)$, and $HC_k(G)$ in line 2.

(3) PHCO directly returns false if the functional safety requirements can't be met in this case; that is, hardware cost optimization failed in lines 3−5.

(4) PHCO iteratively removes ECUs from active ECUs U_{active} in lines 6−18.

(5) PHCO executes an exploratory process for each ECU in U_{active}, sets the u_k state to sleep, and invokes the HEFT on $U_{\text{active}} - \{u_k\}$ to obtain the $R_k(G)$, $RT_k(G)$, and $HC_k(G)$ of G on the active ECUs in lines 7−10.

(6) If no result satisfies functional safety requirements of G, PHCO directly returns true, and the HCO is ends in lines 11−13. Otherwise, PHCO removes the ECU u_{min}, without which $HC_{\text{min}}(G)$ is generated while satisfying functional safety requirements of G in lines 14−15. $HC_{\text{min}}(G)$ can be calculated by

$$HC_{\text{min}}(G) = \min_{u_k \in U_{\text{active}}, RT_k(G) \leqslant RT_{\text{req}}(G), R(G) \geqslant R_{\text{req}}(G)} \{HC_k(G)\}, \qquad (5.5)$$

(7) PHCO updates $R(G)$, $RT(G)$, and $HC(G)$ of the parallel application G in line 16.

EHCO's time complexity is $O(|N|^2 \times |U|^3)$. The details are as follows: (1)the maximum number of times to remove the ECU is $O(|U|)$ times (lines 6−18); (2) all ECUs can be traversed in $O(|U|)$ times (lines 7−10); (3) invoking the HEFT can be done in $O(|N|^2 \times |U|)$ time (line 9).

5.4.3 Example of the PHCO Algorithm

Table 5.4 shows the schedule results of using PHCO algorithm when each ECU is in the sleep state. Each case can meet the real-time requirement because the respective $RT(G)$ is less than 100, but cannot meet the reliability requirement because the respective $R(G)$ is less than 0.95. Hence, no matter which ECU is removed, the functional safety requirements for parallel applications cannot be assured.

Table 5.4: Schedule results using PHCO algorithm.

Sleeping ECU	$R(G)$	$RT(G)$	Whether satisfying functional safety requirements	$HC(G)$
u_1	0.949329	98	No	-
u_2	0.940165	96	No	-
u_3	0.946816	82	No	-

5.5 ENHANCED PROGRESSIVE HARDWARE COST OPTIMIZATION

5.5.1 EPHCO Algorithm

In Table 5.4, we find that reliability values are very close to the reliability requirement of 0.95. If we increase the reliability values without violating the real-time requirement of the parallel application, it is possible to meet the reliability requirement and remove ECUs. In this subsection, we propose EPHCO algorithm as shown in Algorithm 9. The improvement of EPHCO over PHCO is that it introduces the statements in lines 10−15, as described in Algorithm 9:

Algorithm 9 EPHCO Algorithm

Input: $U = \{u_1, u_2, ..., u_{|U|}\}$, G, and $RT_{\text{req}}(G)$, $R_{\text{req}}(G)$
Output: $R(G)$, $RT(G)$, and $HC(G)$

1: $U_{\text{active}} \leftarrow P$;
2: Invoke the HEFT on U_{active} to obtain the initial $RT(G) \leftarrow LB(G)$, $R(G)$, and $HC(G)$;
3: **if** $(RT(G) > RT_{\text{req}}(G)$ or $R(G) < R_{\text{req}}(G))$ **then**
4:     **return** false;
5: **endif**
6: **while** $(U_{\text{active}}$ is not NULL$)$ **do**
7:     **for** (each active ECU u_k in U_{active}) **do**
8:         Set the state of u_k to sleep;
9:         Invoke the HEFT on $U_{\text{active}} - \{u_k\}$ to obtain $R_k(G)$, $RT_k(G)$, and $HC_k(G)$;
10:         **if** $(RT(G) \leqslant RT_{\text{req}}(G)$ and $R_k(G) \leqslant R_{\text{req}}(G))$ **then**
11:             Invoke the RE to increase $R_k(G)$;
12:             **if** $(R_k(G) \geqslant R_{\text{req}}(G))$ **then**
13:                 Re-calculate $HC_k(G)$;
14:             **endif**
15:         **endif**
16:     **endfor**
17:     **if** (no result satisfies $RT_k(G) \leqslant RT_{\text{req}}(G)$ and $R_k(G) \geqslant R_{\text{req}}(G))$ **then**
18:         **return** false;
19:     **else**
20:         Obtain the ECU u_{min} using Eq. (5.5);
21:         $U_{\text{active}} \leftarrow (U_{\text{active}} - \{u_{\text{min}}\})$;
22:         $RT(G) \leftarrow RT_{\text{min}}(G)$, $R(G) \leftarrow R_{\text{min}}(G)$, and $HC(G) \leftarrow HC_{\text{min}}(G)$.
23:     **endif**
24: **endwhile**

---

(1) EPHCO invokes the RE (Algorithm 10) to increase the reliability value $R_k(G)$ after invoking the HEFT if $R_k(G)$ using HEFT is less than the reliability requirement $R_{\text{req}}(G)$ in line 11.

(2) EPHCO re-calculates the $R_k(G)$ after invoking the RE if $R_k(G)$ using RE is larger than or equal to $R_{\text{req}}(G)$ in lines 12–14.

EPHCO algorithm's time complexity is the same as that of the PHCO algorithm (i.e., $O(|N|^2 \times |U|^3)$).

5.5.2 RE Algorithm

RE algorithm's main idea is to migrate each task from the initially assigned ECU to a possible new ECU, which improves the reliability of the parallel application as long as this migration does not violate the priority constraints between tasks and the real-time requirement of the parallel application. The migration is the existence of slack on the ECU that can accommodate the tasks. The details of RE algorithm are described as follows Algorithm 10.

Algorithm 10 RE Algorithm

Input: G, $U_{\text{active}} - \{u_k\}$, given task assignments
Output: $RT_k(G)$, $R_k(G)$,

1: Sort the tasks in a $task\_list$ by ascending order of $rank_u$ values.
2: **while** (**do** $task\_list$ is not NULL)
3:     $n_i \leftarrow task\_list.out()$;
4:     **for** (each ECU u_v in $(U_{\text{active}} - \{u_{\min}\})$) **do**
5:         Calculate $EST(n_i, u_v)$ using Eq. (5.12);
6:         Calculate $LFT(n_i, u_v)$ using Eq. (5.13);
7:         **if** $(LFT(n_i, u_v) - EST(n_i, u_v) < w_{i,v})$ **then**
8:             **continue**;
9:         **endif**
10:         Calculate $R(n_i, u_v)$ using Eq. (2.1);
11:     **endfor**
12:     Select the ECU $u_{\text{pr}(i)}$ with the maximum reliability value for n_i;
13:     $AFT(n_i) \leftarrow RT_{\text{req}}(n_i, u_{\text{pr}(i)})$;
14:     $AST(n_i) \leftarrow \left(RT_{\text{req}}(n_i, u_{\text{pr}(i)}) - w_{i,\text{pr}(i)}\right)$;
15: **endwhile**
16: Calculate $R(G)$ using Eq. (2.2).

RE algorithm's time complexity is $O(|N|^2 \times |U|)$: 1) All tasks can be traversed in $O(|N|)$ times in lines 2−15; 2) All ECUs can be traversed in $O(|U|)$ times in lines 4−11; 3) All precedence and successor tasks of each task need to be traversed to calculate its EST and LFT, and the task can be assigned in any ECU, so it needs $O(|N| \times |U|)$ times in lines 5 and 6.

From the above, it is clear that RE algorithm's time complexity is equal to that of the HEFT, and RE implements low-time complexity reliability enhancement. For the following reasons, EPHCO calls the RE algorithm to improve the reliability values without sacrificing much computational efficiency.

(1) EPHCO has called the HEFT in line 9 (Algorithm 9), and the time complexity of HEFT is the same as RE.

(2) RE and HEFT are not nested in the EPHCO, so adding RE to EPHCO does not increase the EPHCO's time complexity.

(3) RE can skip some ECUs that tasks cannot be inserted into (lines 7−9 in Algorithm 10), so that its computational efficiency is higher than HEFT.

Compared to HEFT, this subsection orders the tasks in the ascending order of $rank_u$ values instead of descending order.

Given the example of u_1 is in the sleep state, the slacks in Fig. 5.2 are between $RT(G)=98$ and $RT_{req}(G)=100$. n_{10} will be optimized first, followed by n_8, n_7, n_9, n_6, n_5, n_2, n_4, n_3, and n_1. We will explain how to calculate the real-time requirement in the following discussion.

5.5.3 Real-Time Requirement of Tasks

We calculate the real-time requirement of the task in this subsection. We first assume that n_{10} has been reassigned in u_2 using the RE shown in Fig. 5.4. We can obtain $AST(n_{10}) = 93$ and $AFT(n_{10}) = 100$. Then we consider that the second task to be reassigned is n_8. In the following discussion, n_8 is used as an example to explain the process.

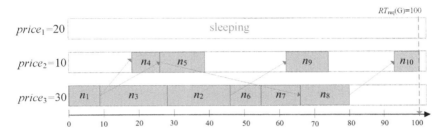

Figure 5.4: Reassignment of n_{10} of the motivational parallel application.

We first move n_8 out of u_3, as shown in Fig. 5.5. Reassigning n_8 to other ECUs makes the reliability of n_8 higher after reassignment and meeting the real-time requirement of n_8 at the same time. Real-time requirement is calculated as follows.

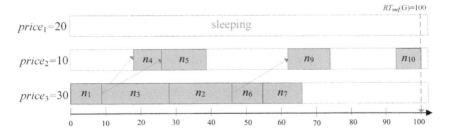

Figure 5.5: Remove n_8 from Fig. 5.4 of the motivational parallel application.

(1) Considering that n_{10} (n_8's successor task) has been reassigned by RE, and it cannot be changed, the LFT of the current task n_i is restricted by its successor tasks because of the precedence constraints among them. LFT is calculated as follows:

$$\begin{cases} LFT(n_{\text{exit}}, u_v) = RT_{\text{req}}(G), \\ LFT(n_i, u_v) = \min_{n_j \in succ(n_i)} \left\{ AST(n_j) - c'_{i,j} \right\}. \end{cases} \tag{5.6}$$

(2) Considering that n_2, n_4, and n_6 (n_8's predecessor tasks) have been assigned by HEFT, and they cannot be changed, the EST of the current task n_i is also restricted by its predecessor tasks owing to the precedence constraints among them. EST is calculated as follows:

$$
\begin{cases}
EST(n_{\text{entry}}, u_v) = 0, \\
EST(n_i, u_v) = \displaystyle\max_{n_h \in pred(n_i)} \left\{ AFT(n_j) - c'_{h,i} \right\}.
\end{cases}
\tag{5.7}
$$

For example, the ESTs and LFTs of n_8 on the ECUs ($U_{\text{active}} - \{u_{\min}\}$) in Fig. 5.5 can be obtained as

$$
\begin{cases}
EST(n_8, u_2) = 70, \\
EST(n_8, u_3) = 55;
\end{cases}
\qquad
\begin{cases}
LFT(n_8, u_2) = 93, \\
LFT(n_8, u_3) = 82.
\end{cases}
\tag{5.8}
$$

(3) Even when the EST and LFT have been obtained on each available ECU for the current task, task reassignment must be further constrained. The reason is that the ECUs in $U_{\text{active}} - \{u_k\}$ are not always available because other tasks have already taken parts of the ECUs and only some slacks remain in the ECUs, as shown in Fig. 5.5. Hence, task reassignment is actually task insertion. The slack set on ECU u_v for n_i is defined as follows:

$$
S_{i,v} = \{ S_{i,v,1}, S_{i,v,2}, ... \},
\tag{5.9}
$$

where $S_{i,v,1}$ denotes the first slack on u_v for n_i. Each slack has a start time (ST) and end time (ET). The v-th slack $S_{i,v,q}$ is defined:

$$
S_{i,v,q} = [t_s(S_{i,v,q}), t_e(S_{i,v,q})],
\tag{5.10}
$$

where $t_s(S_{i,v,q})$ denotes corresponding ST, and $t_e(S_{i,v,q})$ denotes corresponding ET of n_i on u_v. For instance, when reassigning the task n_8 in Fig. 5.5, the slacks on u_2 and u_3 for n_8 are

$$
\begin{cases}
S_{8,2} = \{ [0, 18], [39, 62], [74, 93] \}, \\
S_{8,3} = \{ [66, 100] \}.
\end{cases}
\tag{5.11}
$$

(4) In order to avoid violating the precedence constraints among tasks, the current task n_i should be assigned to the slacks that satisfies the new EST and LFT constraints as follows:

$$
EST(n_i, u_v) = \max \left\{ EST(n_i, u_v), t_s(S_{i,v,t}) \right\},
\tag{5.12}
$$

and

$$
LFT(n_i, u_v) = \min \left\{ LFT(n_i, u_v), t_e(S_{i,v,t}) \right\}.
\tag{5.13}
$$

For instance, the new EST and LFT values of n_8 on all ECUs shown in Fig. 5.5 are as follows:

$$
\begin{cases}
EST(n_8, u_2) = 74, \\
EST(n_8, u_3) = 66;
\end{cases}
\qquad
\begin{cases}
LFT(n_8, u_2) = 93, \\
LFT(n_8, u_3) = 82.
\end{cases}
\tag{5.14}
$$

(5) Considering that the ESTs and LFTs for n_i have been obtained, we can decide which ECU can accept n_i's insertion by judging whether Eq. (5.15) is found:

$$LFT(n_i, u_v) - EST(n_i, u_v) \geqslant w_{i,v}. \qquad (5.15)$$

For example, n_8 can be inserted into u_1 and u_2 because

$$\begin{cases} LFT(n_6, u_2) - AST(n_8, u_2) = 93 - 74 = 19 \geqslant w_{8,2} = 11, \\ LFT(n_8, u_3) - AST(n_8, u_3) = 82 - 66 = 16 \geqslant w_{8,3} = 14. \end{cases} \qquad (5.16)$$

5.5.4 Reliability Enhancement of Tasks

Functional safety requirements of the current task can be obtained in previous sub-sections, the current task can be inserted into ECUs. In this subsection, the strategy of enhancing reliability is as follows: we simply reassign the current task n_i to the insertable ECU with the maximum reliability value without violating the precedence constraints among tasks. That is, the assigned ECU $u_{\mathrm{pr}(i)}$ is determined by

$$R(n_i, u_{\mathrm{pr}(i)}) = \max_{u_v \in (U_{\mathrm{active}} - \{u_k\}), LFT(n_i, u_v) - EST(n_i, u_v) \geqslant w_{i,v}} \{R(n_i, u_v)\}. \qquad (5.17)$$

We assign n_i to $u_{\mathrm{pr}(i)}$ according to AFT and AST of n_i, which is calculated by

$$AFT(n_i) = LFT(n_i, u_{\mathrm{pr}(i)}), \qquad (5.18)$$

and

$$AST(n_i) = AFT(n_i, u_{\mathrm{pr}(i)}) - w_{i,\mathrm{pr}(i)}, \qquad (5.19)$$

respectively. For example, n_8 is inserted into u_2 of Fig. 5.5, because the reliability value of n_8 in u_2 is larger than n_8 in u_3 (Fig. 5.6). The AST and AFT are calculated by Eqs. (5.19) and (5.18) are 82 and 93, respectively.

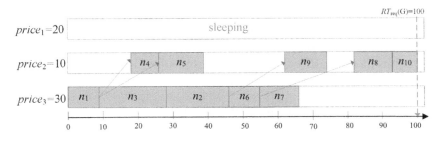

Figure 5.6: n_8 is inserted into u_2 of Fig. 5.5 of the motivational parallel application.

Then, the reliability values of remaining tasks n_7, n_9, n_6, n_5, n_2, n_4, n_3, and n_1 are enhanced using RE. Finally, the task mapping of motivational parallel application on u_2 and u_3 is shown in Fig 5.7, where the reliability value of the parallel application is enhanced to 0.952848.

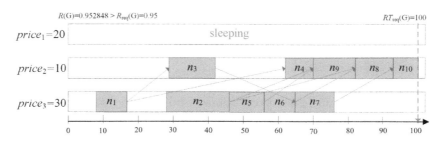

Figure 5.7: RE-generated motivational parallel application's task mapping of the on u_2 and u_3 when u_1 is in the sleep state.

5.5.5 Example of the EPHCO Algorithm

Table 5.5 shows the schedule results by using EPHCO when each ECU is in the sleep state in the first while loop. We can see that when u_1 is in the sleep state, the functional safety requirements can be satisfied, but when u_2 or u_3 is in the sleep state, the reliability requirement cannot be met. Therefore, the functional safety requirements of the parallel application can still be satisfied after removing u_1.

Table 5.5: Schedule results of using EPHCO when each ECU is in the sleep state in the first while loop.

Sleeping ECU	$R(G)$	$RT(G)$	Whether satisfying functional safety requirements	$HC(G)$
u_1	0.952848	92	Yes	40
u_2	0.940165	96	No	-
u_3	0.946816	82	No	-

After the first while loop, Table 5.6 shows the schedule results of using EPHCO when each ECU is in the sleep state in the second while loop. We can see that removing either u_2 or u_3 satisfies neither the reliability requirement of 0.95 nor the real-time requirement of 100. Therefore, merely u_1's sleep in the first while loop is valid, and the task mapping is shown in Fig. 5.7.

Table 5.6: Schedule results of using EPHCO when each ECU is in the sleep state in the second while loop.

Sleeping ECU	$R(G)$	$RT(G)$	Whether satisfying functional safety requirements	$HC(G)$
u_2	0.944405	143	No	-
u_3	0.937067	130	No	-

5.5.6 SEPHCO Algorithm

In spite of the fact that EPHCO can sufficiently remove reasonable ECUs to reduce the hardware cost, it has a high time complexity and thus requires a large computational effort for massively parallel applications. For this reason, we propose a simplified algorithm called SEPHCO in this subsection, as shown in Algorithm 11.

Algorithm 11 SEPHCO Algorithm

Input: $U = \{u_1, u_2, ..., u_{|U|}\}$, G, and $RT_{\text{req}}(G)$, $R_{\text{req}}(G)$

Output: $R(G)$, $RT(G)$, and $HC(G)$

1: Initialize $RT(G) \leftarrow LB(G)$, $HC(G)$, and $R(G)$ by invoking HEFT on U_{active};
2: **if** $(RT(G) > RT_{\text{req}}(G)$ or $R(G) < R_{\text{req}}(G))$ **then**
3:      **return** false;
4: **endif**
5: **for** (each ECU $u_k \in U$) **do**
6:      Set u_k be the sleep state;
7:      Invoke HEFT on $U - \{u_k\}$ to obtain $RT_k(G)$, $R_k(G)$, and $HC_k(G)$;
8:      **if** $(RT(G) \leqslant RT_{\text{req}}(G)$ and $R_k(G) \leqslant R_{\text{req}}(G))$ **then**
9:          Invoke the RE to enhance $R_k(G)$;
10:          **if** $(R_k(G) \geqslant R_{\text{req}}(G))$ **then**
11:              Recalculate $HC_k(G)$;
12:          **endif**
13:      **endif**
14: **endfor**
15: Put all the ECUs into U_{active} according to an ascending of $HC_k(G)$;
16: **while** (U_{active} is not NULL) **do**
17:      Let u_k be the sleep state;
18:      Invoke HEFT on $U_{\text{active}} - \{u_k\}$ to obtain $RT_k(G)$, $R_k(G)$, and $HC_k(G)$;
19:      **if** $(RT(G) \leqslant RT_{\text{req}}(G)$ and $R_k(G) \leqslant R_{\text{req}}(G))$ **then**
20:          Invoke the RE to enhance $R_k(G)$;
21:          **if** $(R_k(G) \geqslant R_{\text{req}}(G))$ **then**
22:              Recalculate $HC_k(G)$;
23:              $U_{\text{active}} \leftarrow (U_{\text{active}} - \{u_k\})$;
24:              $R(G) \leftarrow R_k(G)$, $RT(G) \leftarrow RT_k(G)$, and $HC(G) \leftarrow HC_k(G)$.
25:          **else**
26:              **return** false;
27:          **endif**
28:      **endif**
29: **endwhile**

The simplification of SEPHCO is that it determines the order of ECU removal by attempting to remove each active ECU in lines 5−15, similar to EPHCO. However, instead of redetermining the order of ECU removal in the while loop in lines 16−29, SEPHCO selects the order determined in lines 5−15. By this treatment, the time complexity of SEPHCO is reduced to $O(|N|^2 \times |U|^2)$. Considering that the parallel application has only one valid while loop to select removed ECU, the final hardware cost using the SEPHCO algorithm is also 129.6059 (Fig. 5.7).

The SEPHCO algorithm's time complexity is analyzed in detail as follows: 1) the maximum number of removed ECUs is $|U|$ by invoking the while loop, such that it needs $O(|U|)$ time in lines $16-29$; 2) traversing all ECUs can be done in $O(|U|)$ time in lines $5-14$; 3) invoking the RE and HEFT algorithms can be done in $O(|N|^2 \times |U|)$ time in lines 9 and 11 because they are not nested. Hence, the SEPHCO algorithm's time complexity is $O(|N|^2 \times |U|^2)$.

5.5.7 Optimal Solutions of the Motivational Parallel Application

Given that both PHCO and EPHCO are greedy algorithms and cannot ensure the optimal solution, we try to provide the optimal solution for the motivational parallel application case in this chapter.

(1) Removing two ECUs is not feasible because it does not meet the functional safety requirements of the motivational parallel application.

(2) Considering that the ECU with the highest hardware cost is u_3, if we can find an allocation scheme for tasks that satisfies functional safety requirements in the case of removing u_3, we can get the minimum hardware cost (an optimal solution).

(3) For the motivational parallel application, we can get a total of 40,320 task priority sequences, but only 1680 task priority sequences are valid because most of them violate the priority constraint. n_1 and n_{10} are the highest and lowest task priorities in all the priorities. For each task priority sequence, each task can be executed on u_1 or u_2, so there are 2^{10} task assignment schemes. In other words, there are a total of $680 \times 2^{10} = 1,720,320$ combinations from which we can choose the optimal solution. However, by traversing the above 1,720,320 combinations to calculate their reliability, we do not find a single combination that satisfies the reliability requirement. The maximum reliability of 1,720,320 combinations is 0.946816 ($0.946816 < 0.95$). Fig. 5.8 shows the task assignment on u_1 and u_2 with the maximum reliability when u_3 is in a sleep state, where the task priority sequence is n_1, n_6, n_2, n_5, n_4, n_9, n_3, n_7, n_8, n_{10}, and the response time of the parallel application is 96. Hence, for this parallel application, the hardware cost increases to at least 40, which is already obtained by EPHCO and SEPHCO.

5.6 HARDWARE COST OPTIMIZATION BY CLOSED-TO-OPENED

5.6.1 CEHCO1 Algorithm

We propose CEHCO1 algorithm to optimize hardware cost. Contrary to EPHCO, we set all ECUs to be closed in advance. That is, the CEHCO1 is a closed-to-opened method rather than an opened-to-closed method, as shown in Algorithm 12. We

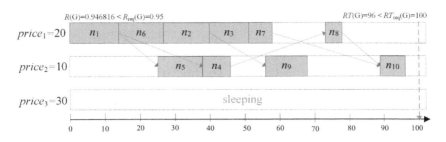

Figure 5.8: Task mapping on u_1 and u_2 with the maximum reliability when set u_3 state to sleep.

Algorithm 12 CEHCO1 Algorithm.

Input: $U = \{u_1, u_2, ..., u_{|U|}\}$, G, and $RT_{req}(G)$, $R_{req}(G)$

Output: $HC(G)$

1: Set all the ECUs to be closed.
2: Sort all the ECUs in U_{closed} according to the descending order of the cost-effectiveness values.
3: Define U_{opened} and its initial value is NULL;
4: **while** (U_{closed} is not NULL) **do**
5:     $u_k \leftarrow U_{closed}.remove()$;
6:     $U_{opened}.add(u_k)$;
7:     Verify whether the functional safety requirements can be satisfied in the opened ECU list U_{opened} by invoking the FFSV1 and FFSV2 algorithms;
8:         **if** (if FFSV1 or FFSV2 returns true) **then**
9:             Calculate $HC(G)$ using Eq. (5.1);
10:             **break**;
11:     **end if**
12: **end while**

---

explain the details of CEHCO1 by combining the motivational parallel application G.

(1) CEHCO1 sets all the ECUs to be closed in line 1, that is, $U_{closed} = \{u_1, u_2, ..., u_{|U|}\}$.

(2) CEHCO1 sorts all the ECUs in U_{closed} according to the descending order of the cost-effectiveness values to provide an opening order of the ECUs' list. The cost-effectiveness is defined as follows.

Definition 5.1. (*Cost-effectiveness*). *The cost-effectiveness of the ECU refers to the reliability performance divided by the unit cost (i.e., an inverse ratio), which can be expressed as*

$$CE(u_k) = \frac{e^{-\lambda_k}}{price_k}. \tag{5.20}$$

In engineering and economics, cost-effectiveness refers to the capability of a product to deliver effectiveness of any sort for its cost. Generally speaking, manufacturers prefer cost-effective products. That is, the opening order of ECUs is: the higher the cost-effectiveness, the higher the priority. Table 5.7 shows the cost-effectiveness values of four ECUs, from which we can obtain the opening order of ECUs, that is, u_1, u_2, u_3, and u_4.

Table 5.7: Cost-effectiveness values of four ECUs.

ECU/Parameter	$CE(u_k)$
u_1	0.100060
u_2	0.050025
u_3	0.03334
u_4	0.02501

(3) CEHCO1 defines the opened ECU list U_{opened}, and its initial value is null in line 3.

(4) CEHCO1 iteratively selects the ECU with the maximum cost-effectiveness to open until the functional safety requirement is satisfied in lines 4−12. The details in the loop are as follows.

(5) CEHCO1 takes out the ECU with the maximum cost-effectiveness from U_{closed} in line 5 and adds this ECU into U_{opened} in line 6.

(6) CEHCO1 verifies whether the functional safety requirements of the parallel application can be assured when executed in U_{opened} by UFFSV in line 7. The UFFSV has been solved in Chapter 2 through presenting the FFSV1 and FFSV2 algorithms (see Chapter 2 for more details about the FFSV1 and FFSV2).

(7) We can obtain a valid hardware cost by using Eq. (5.1) If FFSV1 or FFSV2 returns true in line 9, the loop will be stopped when the functional safety requirements are satisfied (i.e., break in line 10).

5.6.2 Iteration Process of CEHCO1

For the motivational parallel application G, we perform the following iterative process:

(1) We initially open ECU u_1, which has the highest cost-effectiveness. At this point, only one ECU is used to execute the parallel application G. The reliability value is 0.924964 and the response time is 130, as shown in lines 1–2 of Table 5.8. Obviously, either FFSV1 or FFSV2 returns false.

(2) Then, we open u_2, which has the second highest cost-effectiveness. At this point, two ECUs u_1 and u_2 are used to execute the parallel application G. The reliability value is 0.947716 and less than the reliability requirement of 0.95; thus, either FFSV1 or FFSV2 returns false.

(3) We continue to open u_3, which has the third highest cost-effectiveness. At this point, three ECUs u_1, u_2, u_3 are used to execute the parallel application G. The response time value obtained is 80 (less than 95) and reliability value obtained is 0.956380 (larger than 0.95) by invoking FFSV1. The response time value obtained is 97 and reliability value obtained is 0.964544 by invoking FFSV2. Only u_1, u_2, and u_3 are turned on, u_4 is not turned on, FFSV1 and FFSV2 return both true, and the iterative process terminates, at which point the cost is minimal.

Table 5.8: Iteration process using CEHCO1 for the motivational parallel application.

U_{opened}	UFFSV	$R(G)$	$RT(G)$	Satisfying functional safety requirement	$HC(G)$
$\{u_1\}$	FFSV1	0.924964	130	No	10
	FFSV2	0.924964	130	No	10
$\{u_1, u_2\}$	FFSV1	0.947716	94	No	30
	FFSV2	0.947716	94	No	30
$\{u_1, u_2, u_3\}$	FFSV1	0.956380	80	Yes	60
	FFSV2	0.964544	97	Yes	60

Fig. 5.9 shows the task assignment generated by FFSV1 of the parallel application in ECUs $\{u_1, u_2, u_3\}$. The WCRT in this chapter is a theoretical WCRT upper bound rather than the actual communication time. We can find that the functional safety requirements are satisfied and the obtained hardware cost is valid, because the reliability value is 0.956380 >0.95, and the response time value is 80 <97, and $HC(G) = price_1 + price_2 + price_3 = 10 + 20 + 30 = 60$.

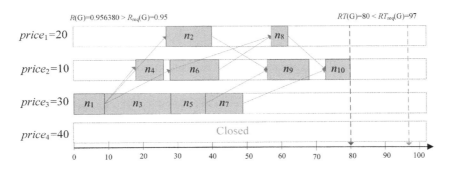

Figure 5.9: Task mapping generated by FFSV1 of the motivational parallel application in ECUs $\{u_1, u_2, u_3\}$.

5.7 HARDWARE COST OPTIMIZATION BY OPENED-TO-CLOSED

5.7.1 CEHCO2 Algorithm

We propose CEHCO2 algorithm to optimize hardware cost, as shown in Algorithm 13. The algorithm description of CEHCO2 is as follows by combining the motivational parallel application G.

Algorithm 13 CEHCO2 Algorithm

Input: $U = \{u_1, u_2, ..., u_{|U|}\}$, G, and $RT_{\text{req}}(G)$, $R_{\text{req}}(G)$, results generated by CEHCO1

Output: $HC(G)$

1: Order all the ECUs in U_{opened} based on the ascending order of the cost-effectiveness values;
2: **while** (there is an ECU that has not been obtained from U_{opened}) **do**
3:    $u_k \leftarrow U_{\text{opened}}.get()$;
4:    Verify whether the functional safety requirements of G can be assured in the $(U_{\text{opened}} - u_k)$ by using the UFFSV algorithm;
5:    **if** (FFSV1 or FFSV2 returns true) **then**
6:      $U_{\text{opened}}.remove(u_k)$;
7:    **endif**
8: **endwhile**
9: Calculate $HC(G)$ using Eq. (5.1).

---

(1) CEHCO2 sorts all the ECUs in U_{opened} based on the ascending order of the cost-effectiveness values in line 1. In contrast to CEHCO1, the opening order of ECUs for CEHCO2 is: the lower the cost-effectiveness, the higher the priority.

(2) Similar to the CEHCO1, only one while loop exists for the CEHCO2 algorithm.

   1) CEHCO2 obtains u_k, it has the lowest cost-effectiveness value from U_{opened}. u_k remains in U_{opened} and has not been removed in line 3.

   2) CEHCO2 verifies whether the functional safety requirements of the motivational parallel application G can be passed in $(U_{\text{opened}} - u_k)$ by using the UFFSV algorithm in line 4.

   3) If FFSV1 or FFSV2 returns true, then CEHCO2 removes u_k from U_{opened} in lines 5−7.

(3) After all the ECUs are obtained from U_{opened} in lines 2−8, we calculate the $HC(G)$ using Eq. (5.1).

5.7.2 Iteration Process of CEHCO2

For the motivational parallel application G, we perform the following iterative process.

(1) We first get u_3, it has the lowest cost-effectiveness value from $U_{\text{opened}} = \{u_1, u_2, u_3\}$. In this case, u_1 and u_2 are used to perform the motivational parallel application G. The response time is $94 < 97$ and reliability value is $0.947716 < 0.95$, as shown in lines $1-2$ of Table 5.9. Thus, the functional safety requirements of the parallel application G cannot be met, and u_3 cannot be removed from U_{opened}.

Table 5.9: Iteration process by using CEHCO2 for the parallel application.

Test ECUs $(U_{\text{opened}} - u_k)$	UFFSV	$R(G)$	$RT(G)$	Satisfying functional safety requirement	U_{opened}	$HC(G)$
$\{u_1, u_2, u_3\}$	FFSV1	0.947716	94	No	$\{u_1, u_2, u_3\}$	60
- u_3	FFSV2	0.947716	94	No	$\{u_1, u_2, u_3\}$	60
$\{u_1, u_2, u_3\}$	FFSV1	0.953897	94	Yes	$\{u_1, u_3\}$	40
- u_2	FFSV2	0.960789	98	No	$\{u_1, u_2, u_3\}$	60
$\{u_1, u_3\}$	FFSV1	0.971999	142	No	$\{u_1, u_3\}$	40
- u_1	FFSV2	0.971999	142	No	$\{u_1, u_3\}$	40

(2) We get u_2 which has the second lowest cost-effectiveness value from $U_{\text{opened}} = \{u_1, u_2, u_3\}$. In this case, u_1 and u_3 are used to perform the motivational parallel application. By invoking FFSV1, the response time is 94, and reliability value is 0.953897, as shown in line 3 of Table 5.9. Thus, the functional safety requirements of the parallel application G can be assured, and u_2 can be removed from U_{opened}, such that U_{opened} is updated to $U_{\text{opened}} = \{u_1, u_3\}$. Notice that the response time is 98, and reliability value is 0.960789 by invoking FFSV2, so the functional safety requirements of the parallel application G cannot be assured in this case. The above results show that the use of UFFSV is more likely to ensure the functional safety requirements than EPHCO with only FFSV2; therefore, it is possible to optimize hardware cost further.

(3) Then, we continue to get u_1 from $U_{\text{opened}} = \{u_1, u_3\}$, and the result shows that u_1 cannot be removed from U_{opened}.

Fig. 5.10 shows the task assignment generated by FFSV1 of the motivational parallel application G in ECUs $\{u_1, u_3\}$. As shown in Fig. 5.10, when reliability value of G is $0.953897 > R_{\text{req}}(G) = 0.95$ and $RT(G) = 94 < RT_{\text{req}}(G) = 97$, the hardware cost of G is valid and $HC(G) = price_1 + price_3 = 10 + 30 = 40$.

5.7.3 CEHCO Algorithm

In view of the parallel application, the final opened ECU set is $U_{\text{opened}} = \{u_1, u_3\}$ after using CEHCO2. However, we can further use CEHCO2 to remove possible ECUs from $U_{\text{opened}} = \{u_1, u_3\}$ without violating the functional safety requirements of the motivational parallel application. Table 5.10 shows the iteration process.

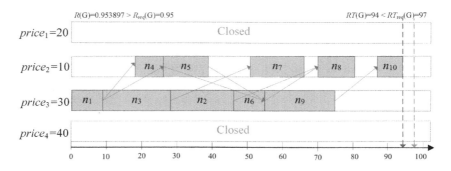

Figure 5.10: Task mapping generated by FFSV1 of the motivational parallel application in ECUs $\{u_1, u_3\}$.

Table 5.10: Iteration process by further using CEHCO2 for the parallel application.

Test ECUs $(U_{opened} - u_k)$	$R(G)$	$RT(G)$	Satisfying functional safety requirement	U_{opened}	$HC(G)$
$\{u_1, u_3\} - u_3$	0.924964	130	No	$\{u_1, u_3\}$	40
$\{u_1, u_3\} - u_1$	0.958295	142	No	$\{u_1, u_3\}$	40

CEHCO2 cannot ensure the functional safety requirements in the case of $\{u_1, u_2, u_3\} - u_3 = \{u_1, u_2\}$ (Table 5.9). However, it may be occurred in CEHCO2 that it meets the functional safety requirements in $\{u_1, u_3\} - u_3 = \{u_1\}$ (Table 5.10). Table 5.10 cannot indicate that the hardware cost is further optimized because the number of ECUs is small for the parallel application G. However, this possibility exists for other parallel applications, because these applications may have many tasks with different WCETs in different ECUs.

The CEHCO algorithm is devised based on the above analysis, as shown in Algorithm 14.

Algorithm 14 CEHCO Algorithm

Input: $U = \{u_1, u_2, ..., u_{|U|}\}$, G, and $RT_{req}(G)$, $R_{req}(G)$
Output: $HC(G)$

1: Invoke CEHCO1 to generate the initial $HC(G)$ of G;
2: **while** (true) **do**
3:     Obtain the $HC_{new}(G)$ of G by invoking CEHCO2;
4:     **if** $(HC_{new}(G) < HC(G))$ **then**
5:         $HC(G) \leftarrow HC_{new}(G)$;
6:     **else**
7:         **break**;
8:     **endif**
9: **endwhile**

---

(1) Init hardware cost $HC(G)$ using CEHCO1 in line 1.

(2) CEHCO obtains the $HC_{\text{new}}(G)$ of the motivational parallel application G by invoking CEHCO2 in the while loop in lines 2−9.

    1) If the $HC_{\text{new}}(G)$ is less than $HC(G)$, then update $HC(G)$ to $HC_{\text{new}}(G)$. An ECU is removed from the opened ECU set by invoking CEHCO2 if the result is $HC_{\text{new}}(G) < HC(G)$.

    2) If the $HC_{\text{new}}(G)$ is equal to $HC(G)$, then the while loop is stopped. That is, the $HC(G)$ reaches a stable value.

5.8 EXPERIMENTS FOR HARDWARE COST OPTIMIZATION ALGORITHMS

5.8.1 Experimental Conditions and Instructions

In this section, we verify the effectiveness of the proposed approach by comparing functional safety metrics (response time value, reliability value) and total hardware cost of a parallel application of the automotive embedded system. We compare PHCO, EPHCO, SEPHCO, and CEHCO algorithms with the IHCO algorithm proposed by Reference [57].

ECU and parallel application parameters are taken from References [73, 74] are as follows: the $w_{i,k}$ and the $c_{i,j}$ are in the range of 10−100 ms, λ_k is in the range of 10^{-6}/ms -9×10^{-6}/ms, and $price_k$ is in the range of \$5−\$100. We also use the hardware cost $HC(G)$, number of opened ECUs $|U_{\text{opened}}|$ for the parallel application, and time consumption of the algorithm as the metrics.

In this section, we use fast Fourier transform (FFT) and Gaussian elimination (GE) parallel applications as experimental objects. Because they have well-defined structures. In particular, FFT is a high parallelism application, and GE is a low parallelism application [58]. By comparing two representative applications with opposite parallelism of the same size, we can fully demonstrate the advantages and characteristics of the proposed algorithm.

All algorithms involve ECU turn-on and turn-off operations; therefore, we analyze the performance of these algorithms with different numbers of ECUs. We set real-time requirement $RT_{\text{req}}(G) = LB(G)$ and reliability requirement $R_{\text{req}}(G) = R_{\text{heft}}(G)$, respectively. $LB(G)$ and $R_{\text{heft}}(G)$ represent lower bound and reliability values generated by HEFT algorithm.

Because our proposed algorithms are aimed at optimizing the hardware cost under functional safety requirements. Therefore, we directly observe the hardware cost, the number of ECUs turned on and the computation time values.

5.8.2 Experimental Details and Analyses

Experiment 1. In this experiment, we observe the hardware costs, and the number of opened ECUs of FFT parallel application with 1151 tasks. We are also concerned about time consumption values of the five algorithms under different numbers of ECUs. The ECU number is changed from 64 to 320 with 64 increments.

Table 5.11 shows the time consumption of all the algorithms. The time consumption of CEHCO includes one call to CEHCO1 and multiple calls to CEHCO2. Compared to other algorithms, EPHCO is quite time-consuming, especially when there are more ECUs. For instance, the time consumption of EPHCO reaches values of 40 and 131 hours when the task numbers are 256 and 320, respectively. IHCO has the best time efficiency in most cases, but IHCO's cost optimization capability is extremely limited. It is of interest that CEHCO produces better time efficiency than SEPHCO, except for $|U| = 64$. When $|U| = 320$, CEHCO requires only 1/2003 of SEPHCO's time to obtain the result. CEHCO is more time efficient than SEPHCO in practice, despite its high time complexity.

Table 5.11: Time consumption of five algorithms when executing FFT parallel application for different numbers of ECUs.

Algorithm	IHCO [57]	EPHCO	SEPHCO	CEHCO1	CEHCO2	CEHCO
$|U| = 64$	2 s	1238 s	48 s	25 s	27 s	82 s
$|U| = 128$	3 s	12199 s (3.4 h)	221 s	31 s	32 s	96 s
$|U| = 192$	8 s	56475 s (15.7 h)	690 s	62 s	50 s	144 s
$|U| = 256$	269 s	144135 s (40 h)	1433 s	88 s	94 s	340 s
$|U| = 320$	21 s	471763 s (131 h)	2635 s	100 s	85 s	232 s

We list the number of iterations of the five algorithms to find out why CEHCO has excellent time efficiency when performing FFT parallel applications for different numbers of ECUs, as shown in Table 5.12. The results show that CEHCO has far fewer iterations than EPHCO. For instance, when $|U| = 320$, CEHCO requires only 100 iterations to obtain the optimized hardware cost of \$951, while EPHCO and SEPHCO require 50799 and 347 iterations to obtain the optimized cost of \$958 and \$18524, respectively. Table 5.13 lists hardware costs the hardware costs.

The sum of the number of iterations of CEHCO1 and CEHCO2 is not always equal to the number of iterations of CEHCO in each column, since CEHCO2 may be called multiple times. Overall, the potential reason for CEHCO's excellent time efficiency is that it requires much fewer iterations than EPHCO and SEPHCO. On further analysis, the essence of the reason for the low number of CEHCO iterations is the use of cost-effectiveness-driven closed-to-opened and opened-to-closed methods: (1) CEHCO1 quickly excludes most cost-effectiveness-driven ECUs (i.e. closed-to-opened); (2) CEHCO2 rapidly excludes some ECUs from the opened ECUs driven by cost-effectiveness (opened-to-closed); and (3) the number of CEHCO2 iterations is decreasing.

Next, we analyze the hardware cost optimization capabilities. Table 5.13 shows that higher costs are generated using IHCO and SEPHCO, while lower costs are generated using EPHCO, CEHCO1, CEHCO2, and CEHCO. In general, IHCO

Table 5.12: Iteration counts of algorithms when executing FFT application for different numbers of ECUs.

Algorithm	IHCO [57]	EPHCO	SEPHCO	CEHCO1	CEHCO2	CEHCO
$|U| = 64$	2	1645	67	35	2	37
$|U| = 128$	1	7661	148	30	1	37
$|U| = 192$	5	18093	218	53	24	77
$|U| = 256$	57	31356	291	66	23	112
$|U| = 320$	1	50799	347	66	34	100

constantly generates the highest hardware costs, followed by SEPHCO. The experimental results show that CEHCO outperforms EPHCO in terms of hardware cost optimization capability. The essence of CEHCO's better hardware cost optimization capability than EPHCO and SEHCO is that CEHCO uses not only FFSV1 but also FFSV2 to increase the likelihood of meeting functional safety requirements, while EPHCO and SEPHCO use only FFSV2, so that CEHCO more thoroughly outperforms EPHCO and SEPHCO in terms of optimizing hardware cost.

Table 5.13: Hardware cost (unit: $) of FFT parallel application for different numbers of ECUs.

Algorithm	IHCO [57]	EPHCO	SEPHCO	CEHCO1	CEHCO2	CEHCO
$|U| = 64$	4430	1363	4133	1657	1543	1443
$|U| = 128$	8604	1243	6579	997	955	955
$|U| = 192$	13054	915	10410	1849	906	906
$|U| = 256$	15412	2102	13581	66	1404	1191
$|U| = 320$	21418	958	18524	2208	951	951

Table 5.14: Number of opened ECUs of FFT parallel application for different numbers of ECUs.

Algorithm	IHCO [57]	EPHCO	SEPHCO	CEHCO1	CEHCO2	CEHCO
$|U| = 64$	63	30	61	34	33	33
$|U| = 128$	128	35	108	30	29	29
$|U| = 192$	188	30	166	53	29	29
$|U| = 256$	238	56	211	66	46	37
$|U| = 320$	320	34	293	66	32	32

In addition to the hardware cost optimization capability, our goal is to determine how many ECUs are turned on to analyze potential causes. The percentage of ECUs opened using the IHCO and SEPHCO algorithms were over 92% and 82%, respectively. IHCO could not even close any ECUs. The number of ECUs opened using EPHCO ranged from 30−56, while the number of ECUs opened using CEHCO ranged from 29−37. Therefore, the number of ECUs turned on with CEHCO is not affected by the total number of ECUs. The result is because CEHCO first identifies the ECUs that must be turned on quickly based on the maximum cost-effectiveness of each ECU, and then quickly identifies the ECUs that must be turned off based on the minimum cost-effectiveness of each ECU.

These experimental results show that CEHCO outperforms SEPHCO and EECHCO in terms of time efficiency and hardware cost optimization capability. The goal of CEHCO is not to trade-off between the time efficiency of SEPHCO and the hardware cost optimization capability of EPHCO, but to achieve both better cost optimization capability and time efficiency than SEPHCO and EPHCO.

Experiment 2. In this experiment, we observe hardware costs, and the number of opened ECUs of the GE parallel application with 1175 tasks. We are also concerned about time consumption values of the 5 algorithms under different numbers of ECUs. Similar to Experiment 1, the ECU number is changed from 64 to 320 with 64 increments.

Table 5.15: Time consumption of algorithms when executing GE application for different numbers of ECUs.

Algorithm	IHCO [57]	EPHCO	SEPHCO	CEHCO1	CEHCO2	CEHCO
$|U| = 64$	2 s	1727 s	21 s	33 s	37 s	100 s
$|U| = 128$	3 s	14164 s (3.9 h)	82 s	26 s	32 s	87 s
$|U| = 192$	11 s	59714 s (16.6 h)	108 s	31 s	33 s	80 s
$|U| = 256$	15 s	166115 s (46 h)	1775 s	200 s	209 s	430 s
$|U| = 320$	29 s	501123 s (129 h)	5536 s	57 s	69 s	148 s

Similar to Table 5.11 for the FFT parallel application, Table 5.15 for the GE parallel application shows that CEHCO is a more time efficient algorithm than SEPHCO for a large number of ECUs. For example, when $|U| = 320$, CEHCO requires only 1/37 times as much as SEPHCO to produce a result. EPHCO is more time consuming, requiring 46 hours and 129 hours to produce results.

Table 5.16 lists iteration counts of algorithms when executing GE parallel application for different numbers of ECUs. Similar to the results in Table 5.12, Table 5.16 also shows that CEHCO needs much fewer iteration counts than EPHCO to obtain superior time efficiency.

The comparison of Tables 5.13 and 5.17 indicates that EPHCO and CEHCO can generate lower hardware costs for the GE parallel application than for the FFT parallel application. Similar to Table 5.13, Table 5.17 shows that the hardware cost difference between EPHCO and CEHCO is small. CEHCO still outperforms EHECO when the numbers of ECUs are 128, 196, 256, and 320. Although CEHCO has a weaker hardware cost optimization advantage than EPHCO for low parallel

Table 5.16: Iteration counts of algorithms when executing GE parallel application for different numbers of ECUs.

Algorithm	IHCO [57]	EPHCO	SEPHCO	CEHCO1	CEHCO2	CEHCO
$|U| = 64$	2	1960	107	34	19	56
$|U| = 128$	4	8201	174	27	12	45
$|U| = 192$	2	18473	276	27	12	39
$|U| = 256$	1	32706	329	114	97	211
$|U| = 320$	2	51224	350	42	28	70

applications, the results show that CEHCO is still strong in hardware cost optimization for the GE parallel applications.

Table 5.17: Hardware cost (unit: $) of GE parallel application for different numbers of ECUs.

Algorithm	IHCO [57]	EPHCO	SEPHCO	CEHCO1	CEHCO2	CEHCO
$|U| = 64$	2	4472	659	1774	703	56
$|U| = 128$	4	8830	428	927	551	427
$|U| = 192$	2	12642	426	522	422	422
$|U| = 256$	1	17127	562	4914	517	517
$|U| = 320$	2	21830	463	1304	461	461

Table 5.18: Number of opened ECUs of GE parallel application for different numbers of ECUs.

Algorithm	IHCO [57]	EPHCO	SEPHCO	CEHCO1	CEHCO2	CEHCO
$|U| = 64$	2	63	16	34	15	14
$|U| = 128$	4	125	11	27	15	12
$|U| = 192$	2	191	14	27	15	15
$|U| = 256$	1	256	20	114	17	17
$|U| = 320$	2	319	17	42	14	14

Furthermore, the trend of the results in Tables 5.17 and 5.18 is generally consistent. The comparison of Table 5.17 and 5.18 shows that EPHCO and CEHCO needs fewer number of ECUs for GE parallel application(low parallelism) than that for FFT parallel application(high parallelism). For example, the number of ECUs opened with EPHCO is only 11 to 20, while the number of ECUs opened with CEHCO is only 12 to 17.

The above analysis shows that EPHCO and CEHCO are more capable of optimizing hardware costs for low parallelism applications than for high-parallelism applications. In addition, CEHCO still maintains excellent time efficiency, while EPHCO is quite time-consuming for a large number of ECUs. Hence, CEHCO outperforms

SEPHCO and EPHCO in terms of time efficiency and hardware cost optimization capability for large-scale heterogeneous distributed embedded systems.

5.9 CONCLUDING REMARKS

In this chapter, we first propose the EHCO algorithm to reduce the ECUs of the embedded system, which the minimum hardware costs can be generated while satisfying functional safety requirements of the parallel application. Considering that partial results generated by EHCO may not satisfy the reliability requirement, we propose a RE algorithm to enhance reliability value of the parallel application, and thereby, we present an EPHCO algorithm to improve the possibility that functional safety requirements are satisfied. On the basis of EPHCO, we devise a SEPHCO algorithm to decrease the complexity of EPHCO.

However, EPHCO has the strong cost-optimization capability but poor time efficiency, while SEPHCO is vice versa in large-scale heterogeneous distributed embedded systems. Therefore, we present a CEHCO approach, which is a combination of CEHCO1 and CEHCO2, to achieve both strong cost optimization capability and superior time efficiency. CEHCO1 iteratively selects the most cost-effective ECU to turn on and overcomes the time efficiency difference (i.e., closed-to-opened). The CE-HCO2 iteration selects the least cost-effective ECU to close, building on CEHCO1 to further optimize hardware cost without losing time efficiency (i.e. opened-to-closed). By significantly reducing the number of iterations, CEHCO overcomes the time inefficiency of the opened-to-closed method; and by employing UFFSV, CEHCO achieves a powerful cost optimization capability.

CHAPTER 6

Development Cost Optimization

I N this chapter, we solve the problem of development cost optimization for parallel applications of embedded systems under ensuring the functional safety requirements based on the ASIL decomposition defined in ISO 26262. We propose two techniques to address the problem. The first technique presents two heuristic algorithms, reliability calculation of scheme(RCS) and minimum development cost with reliability requirement(MDCRR). We first use RCS to calculate the reliability value of each ASIL decomposition scheme; then the MDCRR is used to select the scheme with the minimum development cost without violating reliability constraint. The second technique presents a two-stage solution: (1) functional safety risk assessment and (2) development cost optimization. The first stage assesses the automotive embedded system functional safety risks (including the reliability risk and real-time risk) by proposing the fast risk assessment (FRA) algorithm. "Fast" means short risk assessment time cycle, thereby shorting the development progress. The second stage optimizes the development cost based on the first stage by proposing the dual requirement assurance (DRA) algorithm. "Dual" means that reliability and real-time requirements are considered together. At the end of this chapter, we validate the proposed solutions can not only ensure the automotive embedded system functional safety requirements but also have less development cost of 20%–24% than counterparts by examples and experiments confirmation.

6.1 INTRODUCTION

6.1.1 Development Cost Optimization with Reliability Requirement

The development of safety-critical applications is a highly structured and systematic process as defined by ISO 26262, and as such incur development cost. In general, development cost are the labor used in the system development lifecycle, as opposed to resource costs and hardware costs. Increasing development costs is not a good option, as the automotive industry is a highly cost-sensitive industry for the mass market. A high ASIL task can be decomposed into one or more redundant low ASIL tasks, which is called ASIL decomposition [19, 28, 56]. However, ASIL decomposition affects not only reliability but also development cost.

On the one hand, decomposing a high ASIL task into multiple redundant low ASIL tasks can improve reliability through redundancy and thus easily meet reliability requirement. On the other hand, decomposing a high ASIL task into a low ASIL task can reduce the development cost by 25%−100%. The reason is that the low level has low requirement during the implementation process [19, 56]. However, the decomposition comes at the cost of adding redundant tasks, so in this case the total development cost of the redundant low ASIL tasks is not necessarily lower than the development cost of the original high ASIL tasks.

The development lifecycle of a safety-critical parallel application typically includes analysis, design, implementation, and testing phases. The main goal of the first technique is to minimize the development cost of parallel applications with reliability requirement. The main development cost optimization techniques for embedded systems with reliability requirement are summarized below:

(1) We propose a heuristic algorithm named reliability calculation of scheme (RCS) to calculate the reliability value and development cost of each ASIL decomposition scheme for each task with a low time complexity.

(2) We propose a heuristic algorithm named minimizing development cost with reliability requirement (MDCRR) to select the scheme with the minimum development cost while satisfying the reliability requirement of the parallel application of embedded systems.

6.1.2 Safety Assurance and Development Cost Optimization

ASIL decomposition has both positive and negative impacts on the assurance of functional safety requirements for automotive embedded systems.

First, ASIL decomposition can enhance reliability of the parallel application of the automotive embedded system through a redundant mechanism of tasks. Reliability is one functional safety property in ISO 26262. Therefore, reliability requirement must be assured for a safety-critical automotive embedded system.

Second, ASIL decomposition may cause excessively long end-to-end response time of parallel applications of an automotive embedded system due to the increased redundancy. Given that automotive embedded systems are typical real-time systems, response time is also one important functional safety property. Hence, real-time requirement must be assured for a safety-critical automotive embedded system.

In ISO 26262, malfunctioning behavior is defined as the failure or unintended behavior of the embedded system parallel application with regard to the design intent for this embedded system parallel application [28]. Reliability is related to failure, whereas missing deadline is related to unintended behavior. To achieve system functional safety requirement assurance, both reliability and real-time requirements need to be assured simultaneously. However, it is a conflicting process when increasing reliability for reliability requirement assurance and reducing response time for real-time requirement assurance, such that these two requirements are very hard to be assured simultaneously in practice [20, 63]. Considering the above impacts using ASIL decomposition, optimizing the development cost for parallel applications of an automotive embedded system under ensuring the system functional safety requirements are necessary but challenging.

We propose a technique that simultaneously considers the real-time requirement and reliability requirement to minimize the development cost for parallel applications of embedded systems. The main development cost optimization with reliability requirement and real-time requirement techniques for embedded systems are summarized as follows:

(1) We first assess the automotive system functional safety risks (including the reliability and real-time risk) by proposing the fast risk assessment (FRA) algorithm based on ASIL decomposition. "Fast" means short risk assessment time cycle, thereby shorting the development process.

(2) We optimize the development cost based on the first stage by presenting the dual requirement assurance (DRA) algorithm based on ASIL decomposition. "Dual" means that reliability and real-time requirements are considered together.

(3) The proposed solution can not only ensure the system functional safety requirements but also have less development cost 20%–24% than counterparts by examples and experiments confirmation.

6.2 RELATED WORK

This section mainly summarizes related research from the aspects of development cost, functional safety requirements, and critical level decomposition.

Software development cost optimization is an extensively researched topic. Constructive Cost Model (COCOMO) is one of the most influential software cost models [11]. Researchers have demonstrated the consideration of development cost in the design phase of embedded systems [14]. The problem of deciding on ASIL for parallel applications is usually a manual process performed after hazard analysis and risk assessment. Researchers have begun to propose automated methods to solve this problem [43]. Reference [54] presented a methodology for propagating, transforming, and refining functional safety requirements. Research on SIL decomposition [7, 44] and DAL decomposition [10] are used to reduce development cost. For example, Reference [44] and Reference [7] proposed genetic and search-based metaheuristic algorithms for SIL decomposition. Reference [10] presented a tool named DALculus for the automatic assignment of DALs, such that the smallest DAL may be assigned to an application. The ASIL decomposition for parallel applications was studied recently [19, 56].

In the early development phase, reliability growth becomes critical to support overall development plan decisions, and Reference [35] proposed a multi-objective, multi-stage reliability growth planning approach. Reliability-aware optimization techniques and algorithms typically aim to minimize certain objectives while still satisfying reliability requirement. Higher reliability may lead to longer scheduling lengths (or greater energy consumption) for parallel applications, and the problem of optimizing scheduling lengths (or energy consumption) and reliability is considered a typical bi-objective optimal or Pareto optimal problem [16, 17, 18, 20]. References [36, 38, 79] studied energy-efficient scheduling with a reliability requirement for a parallel application with independent tasks.

Reference [74] proposed a shared recovery-based frequency allocation technique to minimize energy consumption with a reliability requirement for a parallel application on a uniprocessor. The above-mentioned studies were only interested in schedule length or energy consumption minimization with a reliability requirement. Reference [69] presented the DAG_Heu algorithm to minimize the resource cost of parallel applications with real-time requirement and reliability requirements on heterogeneous embedded systems. While MaxRe, RR, and DAG_Heu aim to minimize the resource consumption cost, which is the amount of resources used by the processor when the task is running, the development cost is the labor used in the system development lifecycle. Thus, resource consumption cost and development cost are two completely different concepts. Generally, development cost is reduced by ASIL decomposition in automotive embedded systems [19, 56].

6.3 ASIL DECOMPOSITION

The determination of ASIL is based on the degree of exposure, severity and controllability. Generally, the severity is fixed, because the severity is related to the consequences of the parallel application when the risk occurs, and has been determined in the risk assessment and hazard analysis stages. However, controllability is changeable because it is determined by the current state of the driver. Since this chapter focuses on the design stage, we assume that the parallel application is uncontrollable and is represented by C3 from a conservative point of view. In this chapter, if we want to reduce the ASIL of a parallel application, we need to reduce the exposure.

ISO 26262 provides the ASIL decomposition guide and scheme shown in Fig. 6.1 [28]. Decomposition affects reliability and development cost because tasks have different WCETs and development costs on different ASILs. We consider a task n_i with ASIL C. According to Fig. 6.1(c), we can decompose n_i into two redundant tasks, namely, n_i^1 with ASIL B and n_i^2 with ASIL A. n_i^1 can be further decomposed into two ASIL A tasks according to Fig. 6.1(b). Since all tasks must be certified at the highest criticality level ASIL D for both WCET and development cost, we assume that all tasks of the parallel application are first executed on ASIL D. Then each task is replicated to redundant tasks which can be executed on other ASILs by ASIL decomposition. We can obtain all possible schemes for ASIL D as follows.

(1) ASIL D has three decomposition schemes, ASIL B + ASIL B, ASIL C + ASIL A, and ASIL D, as shown in Fig. 6.1(d).

(2) ASIL C can be decomposed into ASIL B + ASIL A according to Fig. 6.1(c), thus ASIL C + ASIL A can be further decomposed to 2 × ASIL A + ASIL B, as shown the arrow from Figs. 6.2(a) to 6.2(d). Similarly, ASIL B + ASIL

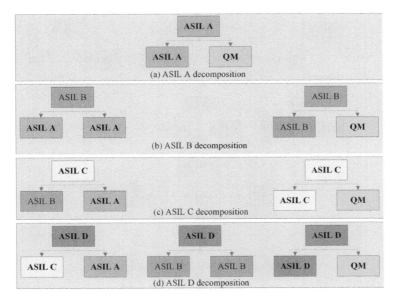

Figure 6.1: ASIL decomposition schemes in ISO 26262 [28].

B can also be further decomposed to ASIL B + 2 × ASIL A, as shown the arrow from Figs. 6.2(b) to 6.2(d). Hence, these two decompositions can only be considered the same result, as shown in Fig. 6.2(d).

(3) ASIL B can be decomposed into ASIL A + ASIL A according to Fig. 6.1(b), such that 2 × ASIL A + ASIL B can be further decomposed to 4 × ASIL A, as indicated by the arrow from Figs. 6.2(d) to 6.2(e).

Finally, five ASIL decomposition schemes of ASIL D are obtained as follows: ASIL C + ASIL A, ASIL B + ASIL B, ASIL D, 2 × ASIL A + ASIL B, and 4 × ASIL A, as indicated by the arrow from Figs. 6.2(a) to 6.2(e).

Our objective is to choose the best scheme for each task.

6.3.1   Exposure and Reliability Requirement

ISO 26262 provides the duration/probability of exposure [28], as shown in Table 3.9. For instance, the probability of E4 is larger than 0.1 of the average running time. In fact, ISO 26262 does not define reliability, but reliability requirements are determined based on exposure levels. For instance, the probability of exposure to E2 is less than 0.01 of the average running time. To ensure functional safety, the actual reliability value must be greater than or equal to $1 - 0.01 = 0.99$, which is considered the reliability requirement in this chapter. The reliability requirements for other exposures can also be calculated according to the above rules. Finally, the reliability requirements for different exposure levels are shown in Table 3.9.

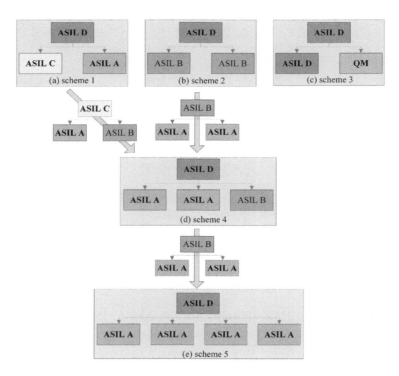

Figure 6.2: ASIL decomposition schemes of ASIL D.

6.4 MODEL AND PROBLEM STATEMENT

Tables 6.1 and 6.2 list the abbreviations and notations that are used in this chapter.

6.4.1 Systems Model

We increased the development costs on the basis of the application model in Section 2.3. A parallel application is represented by a DAG $G=(N, W, M, C, V)$. The details are described as follows:

Table 6.1: Abbreviations in this chapter.

Abbreviation	Definition
DC	Development Cost
MDC	Minimum Development Cost
DAL	Design Assurance level.
RCS	Reliability Calculation of Scheme
MDCRR	Minimum Development Cost with Reliability Requirement
GARR	Genetic Algorithm with Reliability Requirement
FRA	Fast Risk Assessment
DRA	Dual Requirement Assurance

Table 6.2: Notations in this chapter.

Notations	Definition
$w_{i,k,h}$	WCET of the task n_i on the ECU u_k and the ASIL L_h
$R(n_i, u_k, L_h)$	Reliability of the task n_i on the ECU u_k and the ASIL L_h
$R(n_i, scheme_g)$	Reliability of the task n_i with the scheme $scheme_g$
$DC(n_i, L_h)$	Reliability of the task n_i on the ASIL L_h
$DC(G)$	Development cost of the parallel application G
$DC_{\min}(n_i)$	Minimum development cost of the task n_i
$DC_{\max}(n_i)$	Maximum development cost of the task n_i
$DC_{\min}(G)$	Minimum development cost of the parallel application G
$DC_{\max}(G)$	Maximum development cost of the parallel application G

(1) Each task $n_i \in N$ has different WCETs on different ECUs even in the same ASIL. Tasks with high ASIL have a larger WCET than tasks with low ASIL on the same ECU [12].

(2) W is a $4 \times |N| \times |U|$ cube, where $w_{i,k,h}$ represents the WCET of n_i on the ECU u_k and the ASIL L_h.

(3) V is a $4 \times |N|$ matrix, where $v_{i,h}$ represents the development cost of n_i on the ASIL L_h. Each task has a different development cost at different critical levels. Tasks with high ASIL have a greater development cost than tasks with low ASIL [19, 56]. This is because tasks developed under a high ASIL need require more effort to ensure functional safety than those developed under a low ASIL. Similar to References [19, 56], due to the systematic nature of the development process specified by the standard, we assume that the development work for each task on each ASIL is known. In this chapter, we still consider the non-preemptive scheduling of ECUs.

6.4.2 Motivational Example

Fig. 6.3 shows a motivational parallel application with six tasks. Similar to Chapter 2, the arrow between n_1 and n_4 in Fig. 6.3 represents the communication message denoted by $m_{1,4}$, and the WCRT denoted by $c_{1,4} = 14$ if n_1 and n_4 are not assigned to the same ECU.

Table 6.3 lists the WCETs of each task on four different ECUs $\{u_1, u_2, u_3, u_4\}$ and four different ASILs $\{L_A, L_B, L_C, L_D\}$. The weight 7 of n_1, u_2, and L_A in Table 6.3 represents the WCET denoted by $w_{1,2,A}=7$. As mentioned earlier, tasks with high ASIL have greater WCETs than tasks with low ASIL on the same ECU. Due to the heterogeneity of ECUs, tasks with the same ASIL have different WCETs on different ECUs.

Table 6.4 lists the development costs of each task on four different ASILs $\{L_A, L_B, L_C, L_D\}$. The weight 5 of n_1, L_A in Table 6.4 represents the development

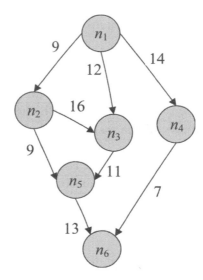

Figure 6.3: A motivational example of a parallel application.

cost denoted by $v_{1,A}=5$. Tasks with high ASIL have larger development costs than that with low ASIL as mentioned earlier.

Table 6.3: WCETs (unit: ms) of all tasks on different ECUs and ASILs.

	n_1				n_2				n_3				n_4				n_5				n_6			
	L_A	L_B	L_C	L_D	L_A	L_B	L_C	L_D	L_A	L_B	L_C	L_D	L_A	L_B	L_C	L_D	L_A	L_B	L_C	L_D	L_A	L_B	L_C	L_D
u_1	4	6	8	10	10	12	14	16	5	7	9	11	14	16	18	20	14	16	186	20	1	3	5	7
u_2	7	9	11	13	9	11	13	15	8	10	12	14	11	13	15	17	12	14	16	18	6	8	10	12
u_3	5	7	9	11	4	6	8	10	13	14	16	18	10	12	14	16	16	18	20	22	3	5	7	9
u_4	8	10	12	14	7	9	11	13	6	8	10	12	6	8	10	12	8	10	12	14	8	10	12	14

Table 6.4: Development costs (unit:kEuro) of all tasks on different ASILs.

	n_1	n_2	n_3	n_4	n_5	n_6
L_A	5	7	5	4	5	8
L_B	8	12	8	7	9	13
L_C	12	17	11	11	14	18
L_D	16	22	14	15	18	22

6.4.3 Development Cost Model

We let $DC(n_i, scheme_g)$ be the development cost of n_i under the ASIL decomposition scheme $scheme_g$. Each scheme has five fixed ASILs shown in Fig. 6.2, then the development cost of each scheme for each task is the sum of those of its copies, as shown in Eq. (6.1). The reason is that multiple copies of the same task developed by these workers substantially spend almost the same development time as explained earlier [19, 56].

$$
\begin{cases}
DC(n_i, scheme_1) = v_{i,\mathrm{C}} + v_{i,\mathrm{A}}, \\
DC(n_i, scheme_2) = 2v_{i,\mathrm{B}}, \\
DC(n_i, scheme_3) = v_{i,\mathrm{D}}, \\
DC(n_i, scheme_4) = 2v_{i,\mathrm{A}} + v_{i,\mathrm{B}}, \\
DC(n_i, scheme_5) = 4v_{i,\mathrm{A}},
\end{cases}
\tag{6.1}
$$

where $v_{i,h}$ represents the development cost of n_i executed in ASIL L_h. Then, the minimum development cost of n_i is

$$
DC_{\min}(n_i) = \min_{g \in [1,5]} DC(n_i, scheme_g),
\tag{6.2}
$$

and the maximum development cost of n_i is

$$
DC_{\max}(n_i) = \max_{g \in [1,5]} DC(n_i, scheme_g).
\tag{6.3}
$$

We let $scheme_{sc(n_i)}$ represent the assigned ASIL decomposition scheme of n_i. Hence, the development cost of the parallel application G is

$$
DC(G) = \sum_{n_i \in N} DC(n_i) = \sum_{n_i \in N} DC(n_i, scheme_{sc(n_i)}).
\tag{6.4}
$$

The minimum development cost of the parallel application G is

$$
DC_{\min}(G) = \sum_{n_i \in N} DC_{\min}(n_i),
\tag{6.5}
$$

and the maximum development cost of the parallel application G is

$$
DC_{\max}(G) = \sum_{n_i \in N} DC_{\max}(n_i).
\tag{6.6}
$$

6.4.4 Reliability Model

The reliability is not defined in ISO 26262, while random hardware failures (including permanent and transient failures) occur unpredictably during the lifecycle of a hardware (ECU) and follows a probability distribution as pointed out in ISO 26262 [28]. We also consider the transient failure (e.g., bit flips) of a task in an embedded system [51, 61].

Let λ_k be the failure rate of ECU u_k, the reliability of n_i executed in u_k and L_h is

$$
R(n_i, u_k, L_h) = e^{-\lambda_k w_{i,k,h}},
\tag{6.7}
$$

and the exposure of n_i executed in u_k and L_h is

$$
E(n_i, u_k, L_h) = 1 - R(n_i, u_k, L_h) = 1 - e^{-\lambda_k w_{i,k,h}}.
\tag{6.8}
$$

We use the probability of hardware failures to get the reliability of the normal execution of a task. The verification of whether a task occurs hardware failures or not requires some measures from the hardware level to resist soft errors and make necessary recovery work.

Each task has $1-4$ copies by performing ASIL decomposition. The copies number of each ASIL decomposition scheme is fixed as pointed in Section 6.3. Let $num(scheme_g)$ be the number of copies of scheme $scheme_g$. For instance, there are three copies for $scheme_4$, as shown in Fig. 6.2. As long as one copy of n_i is successfully completed, then n_i can be successfully completed. Therefore, the reliability value of n_i under the assigned ASIL decomposition scheme $scheme_g$ is

$$
\begin{aligned}
R(n_i, scheme_g) = \\
1 - \prod_{x=1}^{num(scheme_g)} \left(1 - R(n_i^x, u_{\mathrm{pr}(n_i^x)}, L_{cl(\mathrm{pr}(n_i^x))}) \right),
\end{aligned}
\tag{6.9}
$$

where n_i^x is the xth copy of n_i. $u_{\mathrm{pr}(n_i^x)}$ and $L_{cl(n_i^x)}$ are the assigned ECU and ASIL of copy n_i^x, respectively. The difference between Eqs. (6.7) and (6.9) is that the former has the reliability value of n_i in the given ECU and ASIL, whereas the latter has the reliability value of n_i in the given scheme.

The reliability value of the parallel application G is calculated by

$$
R(G) = \prod_{n_i \in N} R(n_i) = \prod_{n_i \in N} R(n_i, scheme_{sc(n_i)}),
\tag{6.10}
$$

where $scheme_{sc(n_i)}$ is the scheme chosen for n_i. We only consider the ECU failures and discard communication failures.

We cannot ensure that all parallel applications of embedded systems are 100% reliable. Hence, so long as the reliability requirement can be assured, then the reliability of the system parallel application is reasonable and meets the standard. However, the reliability requirement is not defined in ISO 26262, but the reliability requirement for given exposure levels can be deduced, as shown in Table 3.9. For example, the reliability requirement for exposure E2 is 0.99.

6.4.5 Problem Statement

We assume that a parallel application is given with a known reliability requirement $R_{\mathrm{req}}(G)$ that would be executed on heterogeneous multiple ECUs set U and ASIL set $\{L_A, L_B, L_C, L_D\}$. The problem to be solved in this chapter is to assign an ECU and ASIL to each task copy while reducing the development cost of the parallel application and meeting its reliability requirements $R_{\mathrm{req}}(G)$. The description is to find the ECU and ASIL assignments for tasks to minimize the development cost of the parallel application G:

$$
DC(G) = \sum_{n_i \in N} DC(n_i),
\tag{6.11}
$$

subject to

$$
R(G) = \prod_{n_i \in N} R\left(n_i, u_{\mathrm{pr}(i)}\right) \geqslant R_{\mathrm{req}}(G),
\tag{6.12}
$$

$R(G)$ represents the reliability value. $DC(n_i)$ represents the development cost of n_i.

6.5   RELIABILITY CALCULATION OF SCHEMES

Since the ASIL D decomposition has fixed schemes as shown in Fig. 6.2, the task in this chapter is to select the scheme and the corresponding ECU that minimizes the development cost while meeting the functional reliability requirements. Therefore, it is first necessary to calculate the possible reliability values of the schemes.

6.5.1   Reliability Calculation

From the above analysis, it is clear that to obtain the reliability value $R(G)$ of a parallel application, we first obtain the allocation scheme of n_i to compute the reliability value of each task (Eq. (6.9)). We list an example of calculating the reliability value of each task on a given scheme. The failure rates of the four ECUs are shown in Table 6.5.

Table 6.5: Failure rates of ECUs $\{u_1, u_2, u_3, u_4\}$.

Parameter	u_1	u_2	u_3	u_4
λ_k	0.01	0.02	0.03	0.04

ASIL D is decomposed into one ASIL C and one ASIL A in $scheme_1$ (Fig. 6.1(a)). Thus, n_1 with $scheme_1$ may be executed on the following ECUs and ASILs:

$$\begin{cases} R(n_1, u_1, L_C) = 0.92311635, \\ R(n_1, u_2, L_C) = 0.80251880, \\ R(n_1, u_3, L_C) = 0.76337949, \\ R(n_1, u_4, L_C) = 0.61878340, \\ R(n_1, u_1, L_A) = 0.96078944, \\ R(n_1, u_2, L_A) = 0.86935824, \\ R(n_1, u_3, L_A) = 0.86070798, \\ R(n_1, u_4, L_A) = 0.72614904. \end{cases} \qquad (6.13)$$

The reliability values of the task in Eq. (6.13) are different because a task has different WCETs on different ECUs or different ASILs, and different failure rates in different ECUs. Since the number of copies in $scheme_1$ is two , we only need to select two of the eight candidates for the reliability value of Eq. (6.13). In order to obtain a high reliability value for this scheme, we use the following steps to select the reliability value of n_1 with $scheme_1$.

(1) We select $R(n_1, u_1, L_A) = 0.96078944$ because it has the maximum reliability value among 8 candidates.

(2) We select the second maximum reliability value for L_C, so $R(n_1, u_1, L_C) = 0.92311635$ can be selected, but u_1 has been occupied by Step (1). We should select the third maximum reliability value $R(n_1, u_2, L_C) = 0.80251880$, where u_2 has not been occupied.

Finally, the reliability of n_1 with $scheme_1$ is as follows calculated by Eq. (6.9):

$$R(n_1, scheme_1) = 1 - (1 - R(n_1, u_1, L_A))\,(1 - R(n_1, u_2, L_C))$$
$$= 1 - (1 - 0.96078944)(1 - 0.80251880) = 0.99225665. \quad (6.14)$$

6.5.2 RCS Algorithm

We present a heuristic RCS algorithm to calculate the reliability of each scheme, as shown in Algorithm 15.

(1) RCS obtains the decomposed ASIL map $map(scheme_g)$ ($< L_\alpha, num_\alpha >, < L_\beta, num_\beta >, ..., < L_\gamma, num_\gamma >$) of $scheme_g$ in line 1.

(2) RCS calculates $R(n_i, u_k, L_h)$ using Eq. (6.8), which is similar to the calculation of Eq. (6.13) in lines 2−6.

(3) RCS orders the reliability values in a *descending_reliability_list* by descending order of $R(n_i, u_k, L_h)$ values in line 7.

(4) RCS calculates $R(n_i, scheme_g)$ based on the heuristic idea to obtain a high reliability value for the scheme as much as possible in lines 8−22.

RCS algorithm's time complexity is analyzed in detail as follows:

(1) Each task's reliability value on each ECU and decomposed ASIL should be calculated in $O(\,|U|)$ time in lines 2−6.

(2) The *descending_reliability_list* should be ordered in $O(\,|N| \times \log|N|)$ time in line 7.

(3) Calculating $R(n_i, scheme_g)$ should traverse all reliability values and verify the occupied ECUs, which can be performed in $O(|U| \times \log|U|)$ time in lines 8−21.

Considering the main time consumption is in line 7, RCS algorithm's time complexity is $O(\,|N| \times \log|N|)$.

6.6 MINIMIZING DEVELOPMENT COST WITH RELIABILITY REQUIREMENT

In this section, we solve the problem of minimizing the development cost of the parallel application of an automotive embedded system under meeting reliability requirement. However, scheduling tasks with functional safety requirements for optimality on multi-ECUs is known to be an NP-hard problem. Many studies use search algorithms to find the decomposition schemes [7, 19, 44, 56], but search algorithms have a large time overhead.

Considering the automotive industry is cost sensitive, it is important to shorten the development lifecycle of parallel applications to reduce development costs. Therefore, from the perspective of development schedule control, using heuristic list scheduling with low time complexity to solve the problem is more suitable than

---

Algorithm 15 RCS Algorithm

Input: $U = \{u_1, u_2, ..., u_{|U|}\}$, $\{L_A, L_B, L_C, ..., L_D\}$, n_i and related values

Output: $R(n_i, scheme_g)$ and related values

1: Obtain the decomposed ASIL map $map(scheme_g)$ ($< L_\alpha, num_\alpha >, < L_\beta, num_\beta >, ..., < L_\gamma, num_\gamma >$) of $scheme_g$, where num_α represents the number of L_α;

2: **for** (each $L_h \in map(scheme_g)$) **do**

3:     **for** (each $u_k \in U$) **do**

4:         Calculate $R(n_i, u_k, L_h)$ using Eq. (6.8);

5:     **endfor**

6: **endfor**

7: Order the reliability values in a *descending_reliability_list* by descending order of $R(n_i, u_k, L_h)$ values.

8: **while** (true) **do**

9:     $R(n_i, u_k, L_h) \leftarrow$ *descending_reliability_list.out*();

10:     **if** (u_k has been occupied by any copy of n_i) **then**

11:         **continue**;

12:     **endif**

13:     **if** (the number of times to select L_h reaches num_h) **then**

14:         **continue**;

15:     **endif**

16:     **if** (the number of times to select any L_h reaches num_h) **then**

17:         **break**;

18:     **endif**

19:     $u_{pr(n_i^x)} \leftarrow u_k$;

20:     $L_{cl(n_i^x)} \leftarrow L_h$;

21: **endwhile**

22: Calculate $R(n_i, scheme_g)$ using Eq. (6.9).

---

search algorithms. In this section, we also use list scheduling to solve thematic problems through task prioritization and task assignment. Task prioritization takes the existing method explained in Section 6.6.1, whereas task assignment is decomposed into two sub-problems: satisfying the reliability requirement and minimizing the development cost, explained in Sections 6.6.2 and 6.6.3, respectively.

6.6.1 Task Prioritization

Similar to References [58, 66, 67], this chapter uses the upward rank value ($rank_u$) of a task given by Eq. (6.15). We use the highest level of ASIL D to determine the task priority as specified by ISO 26262, because a task has different WCETs on different ASILs. Hence, all the tasks are ordered according to the decreasing order of $rank_u$:

$$rank_u(n_i) = \overline{w_{i,D}} + \max_{n_j \in succ(n_i)} \{c_{i,j} + rank_u(n_j)\}. \tag{6.15}$$

$\overline{w_{i,\mathrm{D}}}$ represents the average WCET of task n_i on ASIL D and is calculated by

$$\overline{w_{i,\mathrm{D}}} = \left(\sum_{k=1}^{|U|} w_{i,k,\mathrm{D}} \right) / |U|. \tag{6.16}$$

Table 6.6 shows the upward rank values of all the tasks in Fig. 6.3. We assume that two tasks n_i and n_j meet $rank_{\mathrm{u}}(n_i) > rank_{\mathrm{u}}(n_j)$. If no precedence constraint exists between n_i and n_j, then n_i may not have higher priority than n_j. Finally, the task assignment order in G is $n_1, n_2, n_3, n_5, n_4, n_6$.

Table 6.6: Upward rank values of all tasks in Fig. 6.3 [58].

Task	n_1	n_2	n_3	n_4	n_5	n_6
$rank_{\mathrm{u}}(n_i)$	117	96	67	380	42	10

6.6.2 Satisfying Reliability Requirement

We first traverse all the schemes to obtain the minimum and maximum reliability values. The reliability value of each task on each scheme is obtained by Algorithm 15. And the two values are calculated by

$$R_{\min}(n_i) = \min_{g \in [1,5]} R(n_i, scheme_g), \tag{6.17}$$

and

$$R_{\max}(n_i) = \max_{g \in [1,5]} R(n_i, scheme_g), \tag{6.18}$$

respectively. Product the reliability values of all tasks to get the reliability value of the parallel application G, that is,

$$R_{\min}(G) = \prod_{n_i \in N} R_{\min}(n_i), \tag{6.19}$$

and

$$R_{\max}(G) = \prod_{n_i \in N} R_{\max}(n_i), \tag{6.20}$$

because the reliability of parallel application G is the product of the reliability values of all the tasks (Eq. (6.10)).

According to the above analysis, the reliability requirement $R_{\mathrm{req}}(G)$ should be satisfied during ASIL decomposition. We assume that $R_{\mathrm{req}}(G)$ belongs to the scope $R_{\min}(G)$ and $R_{\max}(G)$,

$$R_{\min}(G) \leqslant R_{\mathrm{req}}(G) \leqslant R_{\max}(G). \tag{6.21}$$

In this chapter, we can calculate the $R_{\min}(G)=0.43171052$ and $R_{\max}(G)=0.99452100$ of the motivational parallel application. We assume that the reliability requirement is $R_{req}(G)=0.9$.

The strategy for meeting the reliability requirement of the parallel application G is shown as follows. We assume that the task to be assigned is $n_{seq(j)}$, where $seq(j)$ represents the j-th assigned sequence. Hence, $\{n_{seq(1)}, n_{seq(2)}, ..., n_{seq(j-1)}\}$ represents the set of tasks that have been assigned, and $\{n_{seq(j+1)}, n_{seq(j+2)}, ..., n_{seq(|N|)}\}$ represents the set of tasks that have not been assigned. To ensure the reliability of the parallel application at each task assignment, we pre-suppose that each task in $\{n_{seq(j+1)}, n_{seq(j+2)}, ..., n_{seq(|N|)}\}$ is assigned to the scheme with the maximum reliability value. Therefore, when assigning $n_{seq(j)}$, the reliability of the parallel application G is calculated by

$$R(G) = \prod_{x=1}^{j-1} R\left(n_{seq(x)}, u_{\mathrm{pr}(seq(x))}\right) \times R\left(n_{seq(j)}\right) \times \prod_{y=j+1}^{|N|} R_{\max}\left(n_{seq(y)}\right). \qquad (6.22)$$

Considering that $R(G)$ should be greater than or equal to $R_{req}(G)$, we have

$$\begin{aligned} R(G) = \prod_{x=1}^{j-1} R\left(n_{seq(x)}, u_{\mathrm{pr}(seq(x))}\right) \times R\left(n_{seq(j)}\right) \\ \times \prod_{y=j+1}^{|N|} R_{\max}\left(n_{seq(y)}\right) \geq R_{req}(G), \end{aligned} \qquad (6.23)$$

namely,

$$R(n_{seq(j)}) \geq \frac{R_{req}(G)}{\prod_{x=1}^{j-1} R(n_{seq(x)}) \times \prod_{y=j+1}^{|N|} R_{\max}(n_{seq(y)})}. \qquad (6.24)$$

Therefore, we let the reliability requirement of task $n_{seq(y)}$ be

$$R_{req}(n_{seq(j)}) = \frac{R_{req}(G)}{\prod_{x=1}^{j-1} R(n_{seq(x)}) \times \prod_{y=j+1}^{|N|} R_{\max}(n_{seq(y)})}. \qquad (6.25)$$

Considering that the minimum reliability of $n_{seq(j)}$ is $R_{\min}(n_{seq(j)})$ (calculated by Eq. (6.17)), $R_{req}(n_{seq(j)})$ should be updated to

$$R_{req}(n_{seq(j)}) = \max\{R_{req}(n_{seq(j)}), R_{\min}(n_{seq(j)})\}. \qquad (6.26)$$

Considering that the maximum reliability of $n_{seq(j)}$ is $R_{\max}(n_{seq(j)})$ (calculated by Eq. (6.18)), $R_{req}(n_{seq(j)})$ should be further updated to

$$R_{req}(n_{seq(j)}) = \min\{R_{req}(n_{seq(j)}), R_{\max}(n_{seq(j)})\}. \qquad (6.27)$$

Hence, the reliability requirement of the parallel application can be transferred to each task. That is, we only let $n_{seq(j)}$ satisfy the following constraint:

$$R(n_{seq(j)}) \geq R_{req}(n_{seq(j)}). \qquad (6.28)$$

Therefore, when assigning task $n_{seq(j)}$, we can directly consider the reliability requirement $R_{req}(n_{seq(j)})$ of $n_{seq(j)}$ and ignore the reliability requirement of the parallel application. Accordingly, a low time complexity heuristic algorithm can be achieved.

6.6.3 Minimizing Development Cost

The minimum development cost that can be obtained is $DC_{\min}(G) = 106$, and its corresponding $R(G)$ is 0.65648591. However, the reliability value cannot reach 0.9, which is the reliability requirement of the motivational parallel application. So, we need to consider that the reliability requirement must be assured in ASIL decomposition. We present an algorithm called MDCRR to reduce the development cost of the parallel application under reliability requirement, as shown in Algorithm 16.

Algorithm 16 MDCRR Algorithm

Input: $U = \{u_1, u_2, ..., u_{|U|}\}$, $\{L_A, L_B, L_C, ..., L_D\}$, G, $R_{\mathrm{req}}(G)$
Output: $R(G)$, $DC(G)$ and related values

1: Sort the tasks in a *descending_task_list* by descending order of $rank_{\mathrm{u}}(n_i)$ values using Eq. (6.15);
2: **while** (*descending_task_list* is not NULL) **do**
3:     $n_i \leftarrow descending\_task\_list.out()$;
4:     **for** $(g \leftarrow 1; g \leq 5; g{+}{+})$ **do**
5:         Calculate $R(n_i, scheme_g)$ using RCS;
6:         Calculate $DC(n_i, scheme_g)$ using Eq. (6.1)
7:     **endfor**
8:     Calculate $R_{\min}(n_i)$ and $R_{\max}(n_i)$ using Eqs. (6.17) and (6.18), respectively;
9:     Calculate $R_{\mathrm{req}}(n_i)$ using Eq. (6.26);
10:    $sc(n_i) \leftarrow 0$, $DC(n_i) \leftarrow \infty$, $R(n_i) \leftarrow 0$;
11:    **for** $(g \leftarrow 1; g \leq 5; g{+}{+})$ **do**
12:        **if** $(R(n_i, scheme_g) < R_{\mathrm{req}}(n_i))$ **then**
13:            continue;
14:        **endif**
15:        **if** $(DC(n_i, scheme_g) < DC(n_i))$ **then**
16:           $sc(n_i) \leftarrow g$;
17:           $R(n_i) \leftarrow R(n_i, scheme_{sc(n_i)})$;
18:           $DC(n_i) \leftarrow DC(n_i, scheme_{sc(n_i)})$;
19:        **endif**
20:        **if** $(DC(n_i, scheme_g) == DC(n_i)$ and $R(n_i, scheme_g) > R(n_i))$ **then**
21:           $sc(n_i) \leftarrow g$;
22:           $R(n_i) \leftarrow R(n_i, scheme_{sc(n_i)})$;
23:           $DC(n_i) \leftarrow DC(n_i, scheme_{sc(n_i)})$;
24:        **endif**
25:     **endfor**
26: **endwhile**
27: Calculate the actual $R(G)$ using Eq. (6.12);
28: Calculate the final $DC(G)$ using Eq. (6.11).

TABLE 6.7 Scheme assignment of the motivational parallel application using the MDCRR algorithm.

n_i	$R_{\text{req}}(n_i)$	$R(n_i, scheme_1)\&$ $DC(n_i, scheme_1)$	$R(n_i, scheme_2)\&$ $DC(n_i, scheme_2)$	$R(n_i, scheme_3)\&$ $DC(n_i, scheme_3)$	$R(n_i, scheme_4)\&$ $DC(n_i, scheme_4)$	$R(n_i, scheme_5)\&$ $DC(n_i, scheme_5)$
n_1	0.90483742	0.99225665 &17	**0.99040688** **&16**	0.90483742 &16	0.99902971 &18	0.99980460 &20
n_2	0.91314970	0.97969496 &24	**0.98137243** **&24**	0.85214379 &22	0.99787492 &26	0.99956709 &28
n_3	0.93161239	**0.98959372** **&16**	0.98774508 &16	0.89583414 &14	0.99802525 &18	0.99950311 &20
n_5	0.93866938	0.96422363 &19	**0.96389111** **&18**	0.81873075 &18	0.99081007 &19	0.99708991 &20
n_4	0.97244400	0.96614003 &15	0.96614855 &14	0.81873075 &15	**0.99293484** **&15**	0.99857324 &16
n_6	0.97933739	**0.99819634** **&26**	0.9958833 &26	0.93239382 &22	0.99987338 &29	0.99997348 &32
		$R(G) = 0.918901856, DC(G) = 115$				

MDCRR attempts to transfer the reliability requirement of the parallel application to each task. Each task selects the solution with the lowest development cost while meeting reliability requirement. The details are as follows:

(1) MDCRR obtains the reliability requirement of each task before it prepares the tasks to be assigned in line 9.

(2) MDCRR skips the solutions that do not meet the reliability requirement in lines 12−14.

(3) MDCRR selects the solution with the minimum development cost for each task while meeting the condition $R(n_i, scheme_g) < R_{\text{req}}(n_i)$ in lines 15−19.

(4) If two solutions have the same development cost, then we select the solution with the higher reliability value in lines 20−24.

6.6.4 Example of MDCRR Algorithm

The process and results of the motivational parallel application using the MDCRR algorithm are illustrated in this subsection. We assume that the constant failure rates for four ECUs are still shown in Table 6.5 and that the reliability requirement is $R_{\text{seq}}(G) = 0.9$. Table 6.7 lists scheme assignment of the motivational parallel application using the MDCRR algorithm. Each row shows the selected scheme and the corresponding reliability value and development cost. For instance, MDCRR selects scheme $scheme_2$ because it has the minimum development cost of 16 in meeting the reliability requirement of 0.90483742. We note that $scheme_3$ also has the minimum development cost of 16, and its reliability value can also meet its reliability requirement. However, if both solutions have the same development cost, the solution with the higher reliability value is chosen. The advantage of this strategy is that the reliability requirement of the remaining tasks can be reduced according to Eq. (6.25). Hence, $R(n_1) = 0.99040688$ and $DC(n_1) = 16$. All the remaining tasks use the same pattern with n_1. Finally, the final reliability value of the parallel application G is $R(G) = 0.918901856$, and the final development cost of the parallel application G is $DC(G) = 115$, which is calculated by Eqs. (6.11) and (6.12), respectively.

The reliability requirement of the parallel application has been met because $R(G) = 0.918901856 > R_{\mathrm{req}}(G) = 0.9$. The minimum development cost of the parallel application is $DC_{\mathrm{min}}(G) = 106$ kEuros (calculated by Eq. (6.5)), which is 9 kEuros less than that using MDCRR. However, the corresponding reliability value for $DC_{\mathrm{min}}(G) = 106$ kEuros is merely $R(G) = 0.65648591$.

Fig. 6.4 shows the task mapping generated by MDCRR of the motivational parallel application G. Different colors of tasks indicate that they are executed on the corresponding ASILs. For instance, n_1 selects the $scheme_2$ containing two ASIL B executed on two ECU u_1 and u_2. The reason is that $scheme_2$ has the minimum development cost of 16 while satisfying n_1's reliability requirement of 0.90483742, as shown in Table 6.7. Finally, the response time is 108 ms.

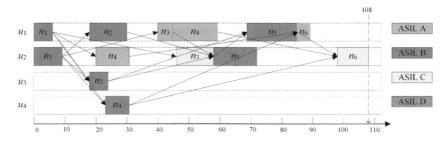

Figure 6.4: Task mapping generated by MDCRR of the motivational parallel application.

6.7 FUNCTIONAL SAFETY RISK ASSESSMENT

We must assess the system functional safety risk before optimizing the development cost. If the system functional safety can be assured, then the following development cost optimization can be conducted; otherwise, it is not necessary to execute the next step to optimize development cost. Hence, the first stage is to assess the embedded system functional safety this section, and then the second stage is to optimize the development cost in the next section.

6.7.1 Reliability Risk Assessment

Given that $scheme_5$ (Fig. 6.2(e)) performs the largest number of redundant copies in four L_A levels, and tasks performed in low ASILs have smaller or equal WCETs than tasks executed in high ASILs, the maximum reliability value must be obtained in $scheme_5$. As long as n_i chooses four ECUs with lowest rate failures, the maximum reliability value of n_i is

$$R_{\mathrm{max}}(n_i) = 1 - \prod_{x=1}^{4} (1 - R(n_i, u_{\mathrm{low\_x}}, L_A)), \tag{6.29}$$

where $u_{\mathrm{low\_1}}$ represents the ECU with the lowest rate of failure, followed by $u_{\mathrm{low\_2}}$, $u_{\mathrm{low\_3}}$, and $u_{\mathrm{low\_4}}$. Next, the maximum reliability value of the parallel application G is

$$R_{\max}(G) = \prod_{n_i \in N} R_{\max}(n_i). \tag{6.30}$$

$R_{\max}(G)$ must be larger than or equal to the reliability requirement $R_{\mathrm{req}}(G)$, namely,

$$R_{\max}(G) \geqslant R_{\mathrm{req}}(G); \tag{6.31}$$

For example, the failure rates of ECUs u_1, u_2, u_3, and u_4 are assumed to be $\lambda_1 = 0.01$, $\lambda_2 = 0.02$, $\lambda_3 = 0.03$, and $\lambda_4 = 0.04$, respectively. Table 6.8 shows the maximum reliability values of the tasks by using Eq. (6.29). The maximum reliability value of the parallel application of an embedded system is $R_{\max}(G) = 0.99968$ using Eq. (6.30). The reliability requirement is assumed to be $R_{\min}(G){=}0.9$. In this case, $R_{\mathrm{req}}(G)$ is reachable and the reliability risk is controllable because

$$0.99968 \geqslant 0.9,$$

according to Eq. (6.31).

Table 6.8: Parameter values of the tasks of the motivational parallel application.

Task	n_1	n_2	n_3	n_4	n_5	n_6
$R_{\max}(n_i)$	0.999992	0.999979	0.999976	0.999934	0.999934	0.999863
$rank_{\mathrm{u}}(n_i)$	117	96	67	38	42	10

6.7.2 Real-Time Risk Assessment

We assess the real-time risk after reliability risk is controllable. Our objective is to judge whether the real-time requirement is met under ensuring the reliability requirement. The formal description is to optimize the response time $RT(G)$, and compare it with the $RT_{\mathrm{req}}(G)$ to judge

$$RT(G) \leqslant RT_{\mathrm{req}}(G), \tag{6.32}$$

under ensuring the reliability requirement:

$$R(G) \geqslant R_{\mathrm{req}}(G). \tag{6.33}$$

Scheduling tasks under the quality of service (QoS) requirements (e.g., real-time requirement or reliability requirement) to achieve optimality (e.g., minimize or maximize) in a multiprocessor (ECU) is NP-hard problem [59]. Hence, obtaining the minimum response time of a parallel application is an NP-hard optimization problem [65]. We also use fast list scheduling to conduct real-time risk assessment.

We adopt the reliability requirement migration strategy to implement fast risk assessment. Each task just chooses the ASIL decomposition scheme and corresponding copies assignment with the minimum EFT under ensuring its reliability requirement.

(1) Reliability requirement of tasks

Two strategies were presented in References [61] and [63] to transfer the reliability requirement of the parallel application to that of each task. Reference [61] solved the problem of optimizing the resource cost under ensuring the reliability requirement by preassigning task's maximum reliability value to each unassigned task without using fault-tolerance. Reference [63] optimizes the redundancy under ensuring the reliability requirement by preassigning task's upper reliability value calculated by Eq. (6.34) to each unassigned task using fault-tolerance.

$$R_{\text{up}}(n_i) = \sqrt[|N|]{R_{\text{req}}(G)}. \tag{6.34}$$

In this section, considering that ASIL decomposition is essentially a special case of fault tolerance, we use the same strategy as Reference [63] to calculate the reliability requirement of tasks is more effective.

The detailed strategy to ensuring the reliability requirement is as follows. Assume that the current task is $n_{s(y)}$ (i.e., the task to be assigned), where $s(y)$ represents the yth assigned task. Two task groups are separated by $s(y)$.

1) The first task group is $\{n_{s(1)}, n_{s(2)}, ..., n_{s(y-1)}\}$, in which the tasks have been assigned.

2) The second task group is $\{n_{s(y+1)}, n_{s(y+2)}, ..., n_{s(|N|)}\}$, in which the tasks have not been assigned.

The tasks are ordered according to the descending order of $rank_{\text{u}}$ values 6.6.1. Table 6.8 shows the upward rank values of all tasks for the motivational parallel application. Finally, the task assignment order in G is $\{n_1, n_2, n_3, n_5, n_4, n_6\}$.

When allocating $n_{s(y)}$, the reliability value of the parallel application G satisfies

$$R_{s(y)}(G) = \prod_{x=1}^{y-1} R\left(n_{s(x)}\right) \times R\left(n_{s(y)}\right) \times \prod_{z=y+1}^{|N|} R_{\text{up}}\left(n_{s(z)}\right) \geqslant R_{\text{req}}(G), \tag{6.35}$$

based on the task's upper reliability value $R_{\text{up}}(n_i)$ in Eq. (6.34). Then, $R\left(n_{s(y)}\right)$ must satisfy

$$R(n_{s(y)}) \geqslant \frac{R_{\text{req}}(G)}{\prod\limits_{x=1}^{y-1} R(n_{s(x)}) \times \prod\limits_{z=y+1}^{|N|} R_{\text{up}}(n_{s(z)})}. \tag{6.36}$$

Then, the reliability requirement of task $n_{s(y)}$ is denoted by

$$R_{\text{req}}(n_{s(y)}) = \frac{R_{\text{req}}(G)}{\prod\limits_{x=1}^{y-1} R(n_{s(x)}) \times \prod\limits_{z=y+1}^{|N|} R_{\text{up}}(n_{s(z)})}. \tag{6.37}$$

The actual reliability value of $n_{s(y)}$ must satisfy

$$R(n_{s(y)}) \geqslant R_{\text{req}}(n_{s(y)}). \tag{6.38}$$

(2) Task assignment

We assign $n_{s(y)}$ to the ECU with the minimum EFT under ensuring $n_{s(y)}$'s reliability requirement $R_{\text{req}}(n_{s(y)})$. Selecting the minimum EFT is a common approach in task assignment [58, 65].

The details of optimizing the response time with ASIL decomposition are as follows: we just optimize the actual finish time of each task by traversing all available ECUs under ensuring its reliability requirement of Eq. (6.37). The assigned ASIL decomposition scheme $scheme_{sc(i)}$ for n_i is determined by

$$AFT(n_i) = EFT(n_i, scheme_{sc(i)}) =$$
$$\min_{g \in [1,5], R(n_i, scheme_g) \geqslant R_{\text{req}}(n_i)} EFT(n_i, scheme_g), \tag{6.39}$$

where $EFT(n_i, scheme_g)$ is calculated by the following:

$$EFT(n_i, scheme_g) = \max_{x \in [1, num(scheme_g)]} EFT(n_i^x, u_{\text{pr}(n_i^x)}). \tag{6.40}$$

The copies of n_i are ordered according to the ascending orders of the ASILs. Because tasks performed in the low ASIL and an ECU have a lower WCET than tasks performed in the high ASIL and that ECU, a locally optimal copy of each scheme is obtained. $R(n_i, scheme_g)$ is calculated by Eq. (6.9).

One possible scenario is as follows. The task assignment in all five ASIL decomposition schemes is based on EFT, which does not ensure the reliability requirement of the task. In this case, instead of pursuing the minimum EFT, we directly consider the maximum reliability value of the scheme calculated by Eq. (6.29). We assign all four copies of the current task with ASIL A to the four ECUs with the lowest failure rate.

6.7.3 FRA Algorithm

In the following, the FRA algorithm of obtaining the response time of the parallel application under ensuring the reliability requirement is proposed, as shown in Algorithm 17.

FRA uses reliability pre-assignment to migrate the reliability requirement of the parallel application to the reliability requirement of each task. And each task simply selects the ASIL decomposition scheme with the minimum EFT and the corresponding copy assignment while ensuring its reliability requirement. The details are described below:

(1) FRA orders all tasks in a *rank_list* by the descending order of $rank_{\text{u}}$ values in line 1.

(2) FRA calculates the reliability requirement of the current task n_i using Eq. (6.37) in line 4.

(3) FRA chooses the ASIL decomposition scheme with the minimum EFT under ensuring $R_{\text{req}}(n_i)$ in line 9. If all the five ASIL decomposition schemes cannot

---

Algorithm 17 FRA Algorithm

Input: $U = \{u_1, u_2, ..., u_{|U|}\}$, $\{L_A, L_B, L_C, L_D\}$, G, $R_{req}(G)$
Output: $RT(G)$

1: Sort the tasks in a $rank\_list$ by descending order of $rank_u(n_i)$ (Eq. (6.15)) values;
2: **while** ($rank\_list$ is not NULL) **do**
3:    $n_i \leftarrow n_{s(y)} \leftarrow task\_list.out()$;
4:    Calculate $R_{req}(n_i)$ using Eq. (6.37);
5:    **for** ($g \leftarrow 1; g \leq 5; g++$) **do**
6:        Calculate $EFT(n_i, scheme_g)$ using Eq. (6.40);
7:        Calculate $R(n_i, scheme_g)$ using Eq. (6.9);
8:    **endfor**
9:    Select the ASIL decomposition scheme with the minimum EFT under assuring $R_{req}(n_i)$;
10:    **if** (all the five schemes cannot satisfy n_i's reliability requirement) **then**
11:        Assign four L_A copies (i.e., $scheme_5$) for n_i to four ECUs with the lowest failure rates;
12:    **endif**
13:    Calculate $AFT(n_i)$ using Eq. (6.39);
14: **endwhile**
15: Calculate $R(G)$ using Eq. (6.10);
16: $RT(G) \leftarrow AFT(n_{exit})$.

---

meet the reliability requirement of n_i, then four copies executed in L_A (i.e., $scheme_5$) are assigned to n_i and executed in four ECUs with the lowest failure rates because this scheme has the maximum reliability value among five schemes in lines 10−12.

(4) FRA calculates the actual reliability and the application finish time in lines 15 and 16.

FRA's time complexity is $O(|N|^2 \times |U|^2)$. In other words, FRA implements low time complexity calculation.

6.7.4 Example of FRA Algorithm

Fig. 6.5 shows task mapping generated by FRA of the motivational parallel application. In this example, n_1, n_3, and n_6 choose $scheme_2$; n_2 chooses $scheme_5$; and n_4 and n_5 choose $scheme_4$. The actual response time is $RT(G)=95$, and

$$RT(G) = 95 \leqslant RT_{req}(G) = 100. \tag{6.41}$$

The actual reliability value of the parallel application is $R(G)=0.906754$ and ensures

$$R(G) = 0.906754 \geqslant R_{req}(G) = 0.9. \tag{6.42}$$

The results of the risk assessment of the motivational parallel application show that functional safety requirements can be assured. In the next section, we perform development cost optimization while ensuring the functional safety requirements.

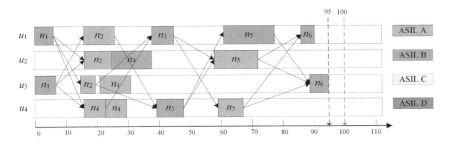

Figure 6.5: Task mapping generated by FRA of the motivational parallel application.

6.8 DEVELOPMENT COST OPTIMIZATION WITH FUNCTIONAL SAFETY REQUIREMENTS

FRA algorithm has ensured the system functional safety requirements (shown in Fig. 6.5), but does not consider the development cost optimization. In this section, we aim to determine the ECU and ASIL assignments of all the copies for each task to optimize the development cost of the parallel application G:

$$DC(G) = \sum_{n_i \in N} DC\left(n_i\right), \tag{6.43}$$

under ensuring the reliability requirement:

$$R(G) = \prod_{n_i \in N} R\left(n_i, u_{\mathrm{pr}(i)}\right) \geqslant R_{\mathrm{req}}(G), \tag{6.44}$$

and ensuring real-time requirement:

$$RT(G) \leqslant RT_{\mathrm{req}}(G). \tag{6.45}$$

We still need to assure both reliability and real-time requirements for each task during in development cost optimization. We first explain how to assure the reliability requirement, and then how to ensure the real-time requirement.

6.8.1 Reliability Requirement Assurance

We first explain the process to ensure the reliability requirement according to the following steps:

(1) Priority of tasks. Different from FRA, where the tasks are prioritized according to the descending order of $rank_\mathrm{u}$ values (i.e., downward optimization from entry to exit tasks), this subsection prioritizes the tasks according to the descending order of $AFT(n_i)$ values generated by FRA (i.e., upward optimization from exit to entry tasks). The reason is that there are some slacks between the FRA-generated $RT(G)$ and the $RT_{\text{req}}(G)$ in some ECUs. For instance, the slacks in Fig. 6.5 are between $RT(G)=95$ and $RT_{\text{req}}(G)=100$. Hence, n_6, n_5, n_3, n_4, n_2, and n_1 will be optimized one by one.

(2)Reliability requirement of tasks. As the initial actual reliability values of all tasks have been known by FRA, the reliability requirement of the current re-assigned task n_i is calculated by

$$R_{\text{req}}(n_i) = \frac{R_{\text{req}}(G)}{\prod\limits_{x=1,x\neq i}^{|N|} R(n_x)}. \tag{6.46}$$

Note that when calculating the reliability requirement of the current task n_i (i.e., $n_{s(y)}$), the reliability values of other tasks cannot be changed.

6.8.2 Real-Time Requirement Assurance

(1) Real-time requirement of tasks. The real-time requirement of the current task n_i is restricted by its successor tasks because of the dependency relationships among tasks,

$$\begin{cases} RT_{\text{req}}(n_{\text{exit}}, u_k) = ET(n_{\text{exit}}, S_k), \\ RT_{\text{req}}(n_i, u_k) = \min \left\{ \min\limits_{n_j \in succ(n_i)} \left\{ AST(n_j) - c'_{i,j} \right\}, ET(n_i, S_k) \right\}, \end{cases} \tag{6.47}$$

where $ET(n_i, S_k)$ is the end time of the available slacks in u_k for n_i. Note that n_i has individual $RT_{\text{req}}(n_i, u_k)$ in different ECUs. Because we aim to re-assign n_i, its previous copy assignments by FRA must be removed.

(2) EST of tasks. Similar to $RT_{\text{req}}(n_i, u_k)$, we must obtain the EST because of the dependency relationships with its predecessors. Each task must have individual $EST(n_i, u_k)$ in different ECUs as follows:

$$\begin{cases} EST(n_{\text{entry}}, u_k) = ST(n_{\text{entry}}, S_k), \\ EST(n_i, u_k) = \max \left\{ \max\limits_{n_h \in pred(n_i)} \left\{ AFT(n_h) + c'_{h,i} \right\}, ST(n_i, S_k) \right\}, \end{cases} \tag{6.48}$$

where $ST(n_i, S_k)$ is the start time of the available slacks in u_k for n_i. For example, the ESTs and RT_{req} of n_6 in all ECUs in Fig. 6.5 are as

$$\begin{cases} EST(n_6, u_1) = 86, \\ EST(n_6, u_2) = 90, \\ EST(n_6, u_3) = 90, \\ EST(n_6, u_4) = 90; \end{cases} \qquad \begin{cases} RT_{\text{req}}(n_6, u_1) = 100, \\ RT_{\text{req}}(n_6, u_2) = 100, \\ RT_{\text{req}}(n_6, u_3) = 100, \\ RT_{\text{req}}(n_6, u_4) = 100. \end{cases} \tag{6.49}$$

(3) Insertable ASIL level (IAL). In view of all the slacks that have been obtained, we can calculate the IAL in each ECU for each task to explore the possible

ASIL decomposition scheme assignment. For example, the IAL values of n_6 in all ECUs are as follows:

$$\begin{cases} IAL(n_6, u_1) = \{L_A, L_B, L_C, L_D\}, \\ IAL(n_6, u_2) = \{L_A, L_B, L_C\}, \\ IAL(n_6, u_3) = \{L_A, L_B, L_C, L_D\}, \\ IAL(n_6, u_4) = \{L_A, L_B\}. \end{cases} \tag{6.50}$$

(4) Reliability of scheme. The reliability value of each possible ASIL decomposition scheme can be calculated based on the IAL. We first ranked IAL in descending order according to the following reliability values below to obtain a high reliability value for a scheme:

$$\begin{cases} R(n_6, u_1, L_A) = 0.960789, \ R(n_6, u_2, L_B) = 0.852144, \ R(n_6, u_3, L_D) = 0.763379, \\ R(n_6, u_1, L_B) = 0.951229, \ R(n_6, u_2, L_C) = 0.818731, \ R(n_6, u_4, L_A) = 0.726149, \\ R(n_6, u_1, L_C) = 0.941765, \ R(n_6, u_3, L_A) = 0.886920, \ R(n_6, u_4, L_B) = 0.670320, \\ R(n_6, u_1, L_D) = 0.932394, \ R(n_6, u_3, L_B) = 0.860708, \\ R(n_6, u_2, L_A) = 0.886920, \ R(n_6, u_3, L_C) = 0.810584. \end{cases} \tag{6.51}$$

Considering the calculation of the reliability value of n_6 with $scheme_4$ in Fig. 6.2(d) (L_A, L_A, and L_B), we do the following search process.

First, we choose $R(n_6, u_1, L_A) = 0.960789$ as it has the maximum reliability value among all L_A candidates.

Second, we choose the second maximum reliability value for L_A, and $R(n_6, u_2, L_A) = 0.886920$ can be chosen.

Third, we choose the maximum reliability value $R(n_6, u_1, L_B) = 0.951229$ among all L_B candidates.

Finally, the reliability value of n_i with $scheme_4$ is $R(n_6, scheme_4) = 0.999382$ calculated by Eq. (6.9).

6.8.3 Optimizing Development Cost

The details of optimizing the development cost are as follows: we just choose the ASIL decomposition scheme with the minimum development cost for each task under ensuring the functional safety requirements. That is, the assigned ASIL decomposition scheme $scheme_{sc(i)}$ and development cost $DC(n_i)$ for n_i are determined by

$$DC(n_i) = DC(n_i, scheme_{sc(i)}) = \min_{g \in [1,5], R(n_i, scheme_g) \geqslant R_{req}(n_i)} DC(n_i, scheme_g)^{.} \tag{6.52}$$

We propose the DRA algorithm as shown in Algorithm 18. Through DRA, the functional safety requirements of each task are simultaneously assured in the process of development cost optimization.

(1) DRA ranks all tasks in a aft_list by the descending order of AFT values in line 1. In the following, all tasks will be traversed.

Algorithm 18 DRA Algorithm

Input: $U = \{u_1, u_2, ..., u_{|U|}\}$, $\{L_A, L_B, L_C, L_D\}$, G, $R_{\text{req}}(G)$, $RT_{\text{req}}(G)$, and FRA-generated assignments

Output: $DC(G)$ and related values

1: Rank the tasks in a aft_task_list by descending order of AFT values using Eq. (6.39);
2: **while** (aft_task_list is not null) **do**
3:     $n_i \leftarrow aft\_task\_list.out()$;
4:     Calculate $R_{\text{req}}(n_i)$ using Eq. (6.46);
5:     Clear the assignments of all the copies of the current task n_i;
6:     Calculate the $EST(n_i, u_k)$ and $RT_{\text{req}}(n_i, u_k)$ of the current task n_i based on individual ECUs;
7:     Calculate the slacks for the current task n_i in individual ECUs;
8:     Calculate the available ASILs of each slack for the current task n_i in individual ECUs;
9:     Rank all the available ASILs for the current task n_i according to the descending order of reliability values;
10:     Calculate the minimum $DC(n_i)$ under assuring n_i's functional safety requirements using Eq. (6.52);
11: **endwhile**
12: Calculate the actual $R(G)$ using Eq. (6.44);
13: Calculate the final $DC(G)$ using Eq. (6.11).

---

(2) DRA calculates $R(G)$ of the current task n_i using Eq. (6.46) in line 4.

(3) DRA calculates $RT(G)$ of the current task n_i using Eq. (6.47), and then obtains the available ASIL decomposition scheme by re-allocation of n_i in lines 5−9.

(4) DRA calculates the minimum development cost under ensuring the functional safety requirements of n_i using Eq. (6.52) in line 10.

(5) DRA calculates the reliability and development cost of the function in lines 12−13.

The time complexity analysis of DRA is the same as FRA and its time complexity is also $O(|N|^2 \times |U|^2)$. Similar to FRA, DRA is suitable from a development progress control perspective.

6.8.4 ■ Example of DRA Algorithm

Fig. 6.6 shows the task mapping of the motivational parallel application G by using DRA. Compared with FRA, DRA further decomposes the ASILs of n_1, n_4, n_3, n_5, and n_6 to optimize the development cost while still ensuring the functional safety requirements. We calculate that the $DC(G)$ in Fig. 6.5 using FRA is 160 kEuros, and it is reduced to 139 kEuros (Fig. 6.6) using DRA.

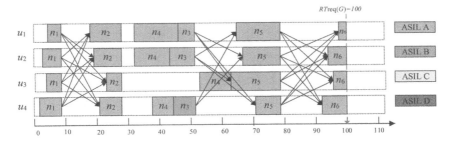

Figure 6.6: Task mapping generated by DRA of the motivational parallel application.

6.9 EXPERIMENTS FOR DEVELOPMENT COST OPTIMIZATION ALGORITHM MDCRR

6.9.1 Experimental Metrics

Given that MDCRR aims to minimize the development cost of the parallel application of an automotive embedded system with a reliability requirement, we choose the actual reliability and the final development cost of the parallel application as the performance metrics used for comparison. Meanwhile, we also consider computation time from a development lifecycle perspective. $DC(G)$ and $R(G)$ are calculated by Eqs. (6.11) and (6.12), respectively.

Algorithms compared with the MDCRR algorithm are the minimum development cost (MDC) and genetic algorithm with reliability requirement (GARR) algorithms. MDC is obtained by Eq. (6.5), and is optimal because it does not consider a reliability requirement or real-time requirement. GARR is a genetic algorithm to search for an enough optimal solution. The Genetic algorithm has been developed to solve the problem of development cost reduction for parallel applications of embedded systems. Hence, the MDC and GARR are suitable as compared algorithms.

6.9.2 Real-Life Parallel Application

We also adopt the real-life parallel application of an automotive embedded system shown in Fig. 2.9 [19]. We also use the same parameter values in Section 2.6.1. The development cost of each task is in the range of 0−30 kEuros.

Experiment 1. In this experiment, we compare the reliability value and the development cost of a real-life parallel application for different reliability requirements. The reliability requirement value is initially set to 0.9 and is increased by 0.1 until 0.99 (in the range of exposure E3 and E2, Table 3.9), with the maximum reliability value of the parallel application being set to 0.999999.

Fig. 6.7 shows the actual reliability of the real-life parallel application for different reliability requirements, and the details are explained as follows.

(1) The minimum reliability value calculated by Eq. (6.19) is 0.753075, and the maximum reliability value calculated by Eq. (6.20) is 0.999999999, respectively.

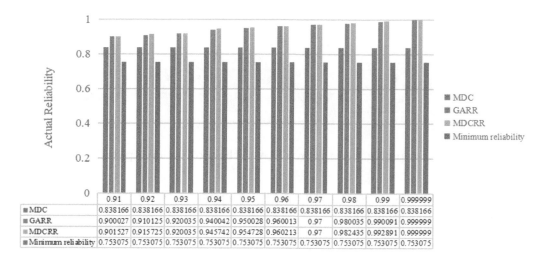

Figure 6.7: Actual reliability of the real-life parallel application for different reliability requirements.

(2) The actual reliability values using the MDCRR and GARR algorithm can always meet and are very close to the corresponding reliability requirements. For MDCRR and GARR, the maximum difference between the actual reliability values and the reliability requirements is only 0.006 and 0.0001, respectively.

(3) In all the cases, the actual reliability values using the MDC algorithm are approximately 0.836166 which cannot satisfy individual reliability requirements.

The above results verify that MDC is not designed to meet the reliability requirements of parallel applications in practice. By contrast, MDCRR can always meet the reliability requirement, and the actual reliability values are close to the reliability requirements without excessive waste.

Table 6.9 shows the development costs of the real-life parallel application for different reliability requirements, and the details are explained as follows.

(1) The minimum and maximum development costs calculated by Eqs. (6.5) and (6.6) are 592 and 901 kEuros, respectively. Thus, the development costs using MDC is 592 kEuros.

(2) We can find that the development costs obtained by MDCRR are not increased linearly with increased reliability requirement. The minimum and maximum development costs are 641 kEuros (R_{req}= 0.9) and 724 kEuros (R_{req} = 0.97), respectively. The reasons are as follows: ASIL decomposition has five fixed schemes that can add more redundant tasks for certain reliability requirements. However, we can select the optimal reliability requirement that has minimum development cost in a given interval. For instance, some automakers may expect a reliability requirement of 0.97, but the actually selected reliability requirement

Table 6.9: Development costs (unit: kEuro) of the real-life application for different reliability requirements.

Reliability Requirement	0.90	0.91	0.92	0.93	0.94	0.95	0.96	0.97	0.98	0.99	0.999999
Maximum development cost	901	901	901	901	901	901	901	901	901	901	901
Development cost using MDC	592	592	592	592	592	592	592	592	592	592	592
Development cost using GARR	603	604	607	610	612	611	618	644	620	621	689
Development cost using MDCRR	641	651	654	655	661	658	662	724	672	678	806

will be 0.98 because it generates the lowest development cost as long as the reliability requirement is not less than 0.97.

(3) The development cost using MDCRR is higher than the minimum development cost using MDC. The reason is that MDCRR needs to meet its reliability requirement and omit some schemes that cannot meet the reliability requirement of the parallel applications in the task assignment (see lines 12−19 of Algorithm 16). However, the development costs using MDCRR are still close to the minimum development costs.

(4) The development costs using GARR are always between those using MDC and MDCRR. The reason is that GARR can find the exact minimum development costs while meeting given reliability requirements, whereas MDCRR is a heuristic list scheduling algorithm to obtain approximate minimum development costs. The results show that GARR can save as much as 14.5% of development cost than MDCRR.

(5) Despite development costs increasing with reliability requirements overall, they do not increase linearly. For example, we can easily observe that when the reliability requirement is 0.97, the second maximum development cost with 724 and 644 kEuros increases for MDCRR and GARR, respectively. Such results indicate that larger reliability requirements do not lead to lower development costs using MDCRR.

Table 6.10: Development costs (unit: kEuro) of synthetic parallel application with 50 tasks for varying reliability requirements.

Reliability Requirement	0.90	0.91	0.92	0.93	0.94	0.95	0.96	0.97	0.98
Maximum development cost	1499	1499	1499	1499	1499	1499	1499	1499	1499
Development cost using MDC	968	968	968	968	968	968	968	968	968
Development cost using MDCRR	1080	1182	1085	1092	1094	1281	1128	1114	1110

Table 6.11: Development costs (unit: kEuro) of synthetic parallel application with 100 tasks for different reliability requirements.

Reliability Requirement	0.90	0.91	0.92	0.93	0.94	0.95	0.96	0.97	0.98
Maximum development cost	2921	2921	2921	2921	2921	2921	2921	2921	2921
Development cost using MDC	1898	1898	1898	1898	1898	1898	1898	1898	1898
Development cost using MDCRR	2110	2223	2135	2136	2328	2426	2152	2160	2168

6.9.3 Synthetic Parallel Application

To further verify the effectiveness of the MDCRR algorithm, we use the synthetic applications with the same real parameter values of the real-life parallel application to observe the results.

Experiment 2. This experiment also indicates the development costs of the parallel applications for different reliability requirements, which are changed from 0.9 to 0.98 with 0.01 increments, because exposure E3 is primarily used in actual system design. The reliability requirements can always be satisfied in all cases in Experiment 1. Therefore, the reliability values of the applications are no longer provided. Tables 6.10 and 6.11 show the development costs of parallel applications with 50 and 100 tasks, respectively, for different reliability requirements. The main details are explained as follows:

(1) The minimum and maximum development costs for the parallel application with 50 tasks are 968 and 1499 kEuros, respectively (Table 6.10), whereas those for the parallel application with 100 tasks are 1898 and 2921 (Table 6.11). Hence, more tasks need more development costs, which are approximately linearly increased.

(2) The development costs for the parallel application with 50 tasks using MDCRR are in the range of 1080–1281 kEuros (Table 6.10), whereas those for the parallel application with 100 tasks are in the range of 1898–2921 kEuros (Table 6.11). Such results further indicate that the development costs are approximately linearly increased with the increment of tasks.

(3) The development costs for the parallel applications with 50 tasks and 100 tasks using MDCRR still do not increase linearly with the increment of reliability requirements. When the reliability requirement is 0.95, both parallel applications need maximum development costs.

In summary, combined with the results of real-life and synthetic parallel applications, the MDCRR is effective in development cost minimization while still ensuring reliability requirement.

6.10 EXPERIMENTAL FOR DEVELOPMENT COST OPTIMIZATION ALGORITHMS FRA AND DRA

6.10.1 Real-Life Parallel Application

In this section, we also adopt the real-life parallel application in an embedded system as same as the last section (Fig. 2.9) from Reference [19] to perform experiments. We also use the same parameter values in Section 2.6.1.

The minimum development cost obtained by Eq. (6.5) is optimal because they do not consider functional safety requirements. In Section 6.6, MDCRR is proposed to optimize the development cost only under ensuring reliability requirement. Hence, the compared algorithms in this section are MDC, MDCRR, FRA, and DRA.

Experiment 1. In this experiment, we compare the development costs of a real-life parallel application of an automotive embedded system under fixed real-time requirement and varying reliability requirements. The reliability requirement value is initially set to 0.9 and is increased by 0.1 until 0.99 (in the range of exposure E3 and E2, Table 3.9), with the maximum reliability value of the parallel application being set to 0.999999 (exposure E1). The real-time requirement is 4333 μs (i.e., $RT_{req}(G) = 4,333\ \mu$s) generated by FRA because it passes the risk assessment. The development costs of the real-life parallel application for different reliability requirements are shown in Fig. 6.8.

(1) MDC always generates the minimum development cost of 641 kEuros. However, MDC cannot ensure the functional safety requirements because its response time value is 7,139 μs (7,139 μs >4,333 μs). The reliability value obtained by MDC is fixed at 0.0944, which is much less than all the reliability requirements.

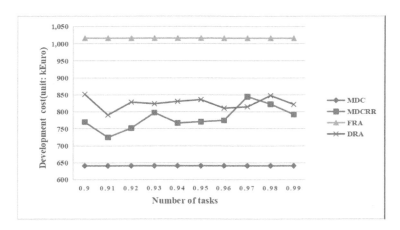

Figure 6.8: Development costs of real-life parallel application for different reliability requirements.

(2) MDCRR, FRA, and DRA can always obtain reliability values larger than or equal to reliability requirements in all cases; thus, they can ensure the reliability requirements.

(3) MDCRR generates fewer development costs than DRA besides for the case that the reliability requirement is 0.97. However, MDCRR obtains very large response time values, such that it cannot ensure the real-time requirement in all cases. The main reason is that MDCRR preassigns maximum reliability value to each unassigned task without using fault-tolerance. This preassignment method is very pessimistic and results in long response time.

(4) FRA always obtains the maximum development cost of 1016 kEuros among the MDCRR, FRA, and DRA algorithms. However, it can ensure the real-time requirement and reliability requirement.

(5) DRA can ensure the real-time and reliability requirements. The actual response time values generated by DRA are equal to the 4333 μs, and the actual reliability values generated by DRA are greater the reliability requirement. Furthermore, the reliability values obtained by DRA are always larger than but quite close to the corresponding reliability requirements. The gap between actual reliability values and reliability requirements is at most 0.008.

(6) The development cost calculated by DRA is not fixed, but decreases as the reliability requirement increases. The development cost varies because the actual reliability value varies as the reliability requirement increases and in this way, the development cost also varies due to the ASIL decomposition. Moreover, the development cost calculated by DRA is between 791 and 852 kEuros. This means that DRA can reduce the development cost by up to 20% compared to FRA under assuring functional safety requirements.

Experiment 2. In this experiment, we compare the development costs of a real-life parallel application with fixed reliability requirement and different real-time requirements. The real-time requirement is changed from $RT_{\mathrm{FRA}}(G)$ to $1.9 \times RT_{\mathrm{FRA}}(G)$ with a $0.1 \times RT_{\mathrm{FRA}}(G)$ increment ($RT_{\mathrm{FRA}}(G) = 4{,}333 \ \mu s$). The reliability requirement is 0.9. In Fig. 6.9, we list the development costs of the real-life parallel application with different real-time requirements.

(1) MDC still generates the development cost of 641 kEuros as shown Fig. 6.8. Because MDC does not care about varying real-time and reliability requirements so MDC's development cost does not change with different real-time or reliability requirements.

(2) The same development costs obtained by MDCRR and FRA are 771 kEuros and 1,016 kEuros, respectively, as shown in Fig. 6.9. Because neither MDCRR nor FRA is concerned with different real-time requirements.

(3) The development costs obtained by DRA are not fixed and reduced with increased real-time requirement. The development costs are reduced from 852 to 681 kEuros. With the increased real-time requirement, more slacks are available. Hence, more ASIL decomposition schemes can insert these slacks, and there are more possibilities to choose the ASIL decomposition scheme with the minimum development cost as expressed in Eq. (6.52).

(4) When the real-time requirement is larger than or equal to $6499.5 \mu s$, DRA obtains a lower development cost than MDCRR. The experiment results show that DRA is very effective in development cost optimization for relaxed real-time requirements.

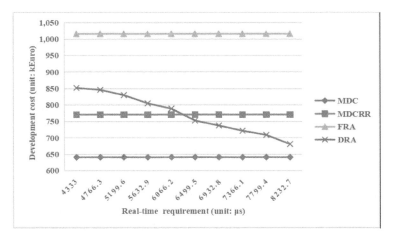

Figure 6.9: Development costs of real-life parallel application for different real-time requirements.

6.10.2 Synthetic Parallel Application

Experiment 3. This experiment shows the development costs of synthetic parallel applications for various numbers of tasks. Synthetic parallel applications can be generated by the task graph generator [1]. According to the parameter setting requirements of the synthetic parallel application, the communication rate is set to 1, the shape parameter is set to 1, and the heterogeneity factor is set to 0.5. More details and explanations about parameter setting can be found in Section 6.9. Task numbers of synthetic applications are also changed from 50 to 100 with 10 increments. The real-time requirement is fixed at $RT_{\text{FRA}}(G)$ obtained by FRA, and the reliability requirement is fixed at 0.9 for parallel applications. The development costs of synthetic parallel applications for various numbers of tasks are shown in Fig. 6.10.

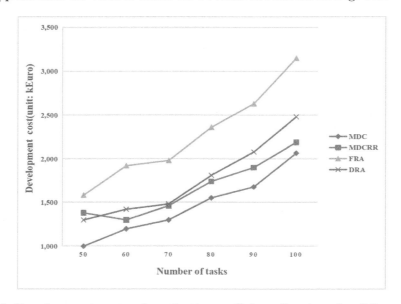

Figure 6.10: Development costs of synthetic parallel applications for different numbers of tasks.

(1) The development costs using all the approaches are increased with the increase of task numbers. The reason is that the development cost of a parallel application is the sum of those of all tasks, such that more tasks will result in larger development costs.

(2) The MDC and FRA approach still obtain the minimum and maximum development costs, respectively. However, the reliability values generated by MDC are in the range of 0.06−0.22, which is much less than the reliability requirement of 0.9.

(3) MDCRR has lower development costs than DRA in most cases and the maximum difference is 120 kEuros, but the MDCRR cannot ensure the functional safety requirements.

(4) DRA has lower development costs than FRA in all cases, and DRA can reduce the development costs by as much as 20%−24% compared with FRA under ensuring the functional safety requirements.

In summary, DRA shows better results for real-life and synthetic parallel applications than other counterparts under ensuring the functional safety requirements.

6.11 CONCLUDING REMARKS

This chapter develops effective development cost minimization solutions for a parallel application with the functional safety requirements of an automotive embedded system by using ASIL decomposition during the design phase. This chapter presented two techniques to solve the problem. In the first technique, RCS calculates the reliability value of each ASIL decomposition scheme, and MDCRR transfers the reliability requirement of the parallel application to each task and selects the ASIL decomposition scheme with the minimum development cost while satisfying the reliability requirement of the task. MDCRR is validated with real-life and synthetic parallel application in various situations.

Given that the real-time requirement is also one of the important functional safety requirements in real-time embedded systems, our studies need simultaneously consider the real-time and reliability requirements to minimize the development cost for parallel applications on embedded systems. It is feasible to consider timing requirements and reliability requirements one by one in the design phase, or to consider them simultaneously to conduct bi-objective optimization.

Therefore, in the second technique, we solved the problem of development cost optimization for a parallel application of embedded system simultaneously ensuring the real-timing requirement and reliability requirement by proposing the FRA and DRA algorithms. FRA is a fast algorithm for embedded system functional safety risk assessment, which decides the feasibility of the development cost optimization. DRA ensures dual requirements (i.e., reliability and real-time requirements) simultaneously and can still ensure the functional safety requirements in the process of development cost optimization. DRA shows quite a good ability for automotive development cost optimization for real-life and synthetic automotive embedded system applications.

CHAPTER 7

Summary and Future Research

7.1 SUMMARY

This book focuses on functional safety for embedded systems. In this book, the parallel application model of embedded systems is described by a DAG.

In Chapter 2, we present a fast functional safety verification approach for the parallel application of an automotive embedded system. First, we propose the FFSV1 algorithm to find the solution of the minimum response time while meeting the reliability requirement. Second, we propose the FFSV2 algorithm to find the solution of maximum reliability while meeting the response time requirement. Based on the above two algorithms, we further propose UFFSV, which is an effective combination of FFSV1 and FFSV2 algorithms. UFFSV is a fast heuristic method that can shorten the lifecycle of application development.

In Chapter 3, we propose the reliability enhancement method SSFSE to enhance functional safety. SSFSE method combines the HEFT, BFSE algorithms and the FFSE, RBFSE, and RFFSE algorithms introduced in Chapter 3. SSFSE method enhances safety by using stable stopping methods based on forward and backward recovery through primary and backup repetition. SSFSE can bring the exposure level down from E3 to E1 toward a higher level of safety.

In Chapter 4, we proposed the GMNRA and GMFRA algorithms, in which reliability values based on geometric mean are pre-assigned to unassigned tasks. The geometric mean allows the pre-assigned reliability values of unassigned tasks to converge, thus allowing a more balanced distribution of reliability requirements. GMNRA and GMFRA can effectively reduce the response time of automotive functions while ensuring their reliability requirement.

In Chapter 5, we present two hardware cost optimization techniques for parallel applications in automotive embedded systems. In the first technique, we propose EHCO, EEHCO, and SEEHCO algorithms to reduce the number of ECUs on the premise of meeting the functional safety requirements. However, the time efficiency of the first technique is general in large-scale distributed embedded systems. Therefore, we present a cost-effectiveness-driven hardware cost optimization

DOI: 10.1201/9781003391517-7

technique (CEHCO), which combines CEHCO1 and CEHCO2 algorithms, to achieve powerful hardware cost optimization capability and superior time efficiency simultaneously.

In Chapter 6, we first solve the problem of the development cost minimization for the parallel application of an automotive embedded system while satisfying its reliability requirement during the design phase by proposing two heuristic algorithms, RCS and MDCRG. RCS calculates the reliability value of each ASIL decomposition scheme, and MDCRG transfers the reliability requirement of the parallel application to each task and selects the ASIL decomposition scheme with the minimum development cost while satisfying the reliability requirement of the task. We further solve the problem of development cost optimization for an end-to-end embedded system application under simultaneously ensuring the timing constraint and reliability requirement in embedded system application by proposing the FRA and DRA algorithms. FRA is a fast algorithm for embedded system functional safety risk assessment, which decides the feasibility of the development cost optimization. DRA ensures dual requirements (i.e., reliability and real-time requirements) simultaneously and can still ensure the functional safety requirements in the process of development cost optimization.

7.2 FUTURE RESEARCH

With the gradual development of electrification, intelligence, networking, and sharing of automobiles, as well as the gradual improvement of social infrastructure and automobile industry ecology, self-driving vehicle is considered to be the highest point of future automobile industry development and one of the current global innovation hotspots. The future trend of automotive functional safety design methods will directly face the self-driving vehicle and the next functional safety standards. In order to expand future research, the following factors should be considered.

SAE Levels of Driving Automation

In self-driving vehicles, systems driven entirely by humans are evolving into human-machine driven systems. In 2019, SAE J3016, the SAE level of driving automation standard(i.e., classification and definition of terms related to driver automation systems for road motor vehicles [49]), updated the J3016 level of autonomous driving graphics to reflect the evolving standard [52]. The SAE level clearly describes and defines the content and proportion of the vehicle engine system and driver's work. However, there may be a deviation in the actual execution of HARA by the ASIL of a self-driving application. Controllability in ASIL refers to the controllability of the driver state. Specifically, when the operation state of the self-driving application is combined with the state of the driver himself, the auxiliary driving behavior of human-machine interaction perception will be formed. These assistant driving behaviors sense and integrate the state of the driver and vehicle engine system, which can effectively improve the safety of self-driving, but may also lead to the instability of the self-driving state of the vehicle.

Safety of the Intended Functionality

Functional safety problems of nonautonomous vehicles arise from system failure (e.g., using unsafe real-time analysis approaches) or random hardware failure (e.g., using unsafe reliability analysis approaches). The self-driving vehicle may have uncertain outputs because they use artificial intelligence (AI) algorithms. This uncertainty is an unexpected factor that can lead to functional deviations or even safety accidents in automotive applications [6, 40]. Overall, the impact of AI algorithms on functional safety design is as follows: (1) AI algorithms generate unpredictable system failures; (2) AI algorithms have an uncertain impact on system behavior; (3) AI algorithms evolve with the development of data processing; and (4) AI algorithms have vulnerabilities and risks.

Because the output of the image recognition algorithm based on artificial intelligence is affected by multiple factors, the output results are uncertain, which will lead to unpredictable safety risks (i.e., in different driving environments, the same obstacle may be recognized as different objects). Since the world's first self-driving vehicle accident in 2016 [21], self-driving vehicles' safety accidents have frequently occurred on many brands of vehicles, such as Waymo, Uber, and Tesla. The causes of these safety accidents include but are not limited to the sudden failures of the self-driving system, the design defects of intelligent applications, and the AI algorithms that lead to inaccurate pedestrian recognition. The above-mentioned accidents demonstrate that it is urgent to build vehicle functional safety standards and design methods for self-driving vehicles. Safety is always the most important issue for self-driving vehicles. Any accident involving functional safety can weaken users' trust in self-driving vehicles, and even hinder the development and prosperity of self-driving vehicles.

Safety of the intended functionality (SOTIF) means that there is no unreasonable risk due to inadequate functionality of the intended function or reasonably foreseeable misuse by personnel [29]. To avoid unreasonable risk due to the above unintended factors, the publicly available specification (PAS) SOTIF for self-driving cars (i.e. ISO/PAS 21448) has been released in January 2019 and the official version is expected to be released in 2022. The standard focuses on SAE automation levels 1 and 2 driver assistance functions. Compliance with SOTIF design becomes a prerequisite for the practical use of self-driving vehicles. However, how AI algorithms and technologies will affect safety analysis, safety mechanisms, verification and validation has become a pressing research issue. Furthermore, how to analyze the possible interactions between functional safety and SOTIF is also a question worth exploring.

Automotive Cyber Security

For a long time in the past, suppliers of vehicle components and vehicle manufacturers focused primarily on functional safety to ensure the safety of drivers and passengers. Nowadays, with the development of terminal information technology and the increase of wireless interfaces in vehicles, the number of network interfaces that can be attacked by automotive embedded systems has gradually increased. Attackers can launch cyber attacks through terminals to prevent the normal operation of automotive embedded systems, causing automotive system failures and even endangering death, injury, and harm to drivers, passengers, and roadside pedestrians. Hence, security measures have to be taken to ensure that automotive embedded systems are well resilient to network attacks. To this end, SAE officially released the cyber security guidebook for electronic vehicle systems (SAE J3061) in January 2016 [50].

In fact, functional safety and cyber security can affect each other. On the one hand, the introduction of cyber security protection strategy (including encryption and decryption, and increase in message authentication codes (MAC)) causes communication delay overhead, which affects reliability and response time (i.e., functional safety) [31]. On the other hand, the redundancy strategy of functional safety enhancement may introduce additional illegal attack interfaces, which will increase the success probability of attackers. Therefore, the second edition of ISO 26262 regards cyber security as an important part of functional safety [30]. The new version of ISO 26262 stipulates that it is desirable to establish a relationship between functional safety and cyber security such that the impact of each can be clearly analyzed. In order to better enable interested parties to understand relevant specifications, the second edition of ISO 26262 provides guidance on potential interactions between functional safety and cyber security in Appendix E, Part II. In addition, the automotive cyber security standard ISO/SAE AWI 21434 (i.e., road vehicle cyber security engineering) was released in 2020 [5].

In recent years, the interaction between the integrity of cyber security and the real-time performance of functional safety has made some progress in the design and research of automotive embedded systems based on CAN bus. It is an effective method to resist masquerade attacks by adding MAC to the message payload. References [37, 78] focused on safety and security requirements assurance. Reference [68] reduced the bandwidth utilization of CAN with flexible data rate (CAN FD) messages to the greatest extent by adding a MAC with a fixed size of 4B to ensure safety and security requirements. Reference [70] presented to use of MAC with fixed size and timestamp to achieve safety and security assurance. Reference [34] improved the effectiveness and quality of system control by adding MAC while assuring the real-time requirements of functional safety. In addition, with the gradual promotion of TSN in the design of self-driving vehicles, security-aware routing and scheduling for the real-time parallel application of automotive embedded systems have become a new trend [39].

Bibliography

[1] https://sourceforge.net/projects/taskgraphgen/. 2015.

[2] Real-time computing - wikipedia. Apr. 2014.

[3] Work-related accidents and injuries cost eur 476 billion a year according to new global estimates. Apr. 2017.

[4] The case study of automotive airbag reliability. pages 1–5, Oct. 2012.

[5] Iso/sae awi 21434 road vehicles-cybersecurity engineering, status overview. Oct. 2017.

[6] S. Ali and T. Yue. U-test: Evolving, modelling and testing realistic uncertain behaviours of cyber-physical systems. In *IEEE International Conference on Software Testing*, 2015.

[7] Luis Silva Azevedo, David Parker, Martin Walker, Yiannis Papadopoulos, and Rui Esteves Araujo. Automatic decomposition of safety integrity levels: optimization by tabu search. In *Proc. 2nd Workshop Critical Automotive Applications: Robustness amd Safety) of the 32nd Int. Conf. Computer Safety, Reliability and Security*, page NA, 2013.

[8] A. Benoit, M. Hakem, and Y. Robert. Fault tolerant scheduling of precedence task graphs on heterogeneous platforms. In *Parallel and Distributed Processing, 2008. IPDPS 2008. IEEE International Symposium on*, 2008.

[9] Anne Benoit, Fanny Dufossé, Alain Girault, and Yves Robert. Reliability and performance optimization of pipelined real-time systems. *Journal of Parallel and Distributed Computing*, 73(6):851–865, June 2013.

[10] Pierre Bieber, Rémi Delmas, and Christel Seguin. Dalculus–theory and tool for development assurance level allocation. In *Proc. Int. Conf. Computer Safety, Reliability, and Security*, pages 43–56. Springer, 2011.

[11] Barry Boehm, Chris Abts, and Sunita Chulani. Software development cost estimation approaches—a survey. *Annals of Software Engineering*, 10(1-4):177–205, Nov. 2000.

[12] A. Di Burns and R. Davis. Mixed-criticality systems: A review (eighth edition). *Website. http://www-users.cs.york.ac.uk/burns/review.pdf*, 2016.

[13] Robert I Davis, Sebastian Altmeyer, and Alan Burns. Mixed criticality systems with varying context switch costs. In *2018 IEEE Real-Time and Embedded Technology and Applications Symposium (RTAS)*, pages 140–151. IEEE, 2018.

[14] James A Debardelaben, Vijay K Madisetti, and Anthony J Gadient. Incorporating cost modeling in embedded-system design. *IEEE Design and Test of Computers*, 14(3):24–35, Jul. 1997.

[15] Marco Di Natale, Haibo Zeng, Paolo Giusto, and Arkadeb Ghosal. *Understanding and using the controller area network communication protocol: theory and practice*. Springer Science and Business Media, 2012.

[16] Atakan Dogan and Füsün Ozguner. Matching and scheduling algorithms for minimizing execution time and failure probability of applications in heterogeneous computing. *IEEE Trans. Parallel Distrib. Syst.*, 13(3):308–323, Mar. 2002.

[17] Atakan Doğan and Füsun Özgüner. Biobjective scheduling algorithms for execution time–reliability trade-off in heterogeneous computing systems. *The Computer Journal*, 48(3):300–314, Mar. 2005.

[18] Jack J Dongarra, Emmanuel Jeannot, Erik Saule, and Zhiao Shi. Bi-objective scheduling algorithms for optimizing makespan and reliability on heterogeneous systems. In *Proc. 19th ACM Int. Symp. on Parallel algorithms and architectures*, pages 280–288. ACM, 2007.

[19] Junhe Gan, Paul Pop, and Jan Madsen. *Tradeoff Analysis for Dependable Real-Time Embedded Systems during the Early Design Phases*. PhD thesis, Technical University of DenmarkDanmarks Tekniske Universitet, Department of Informatics and Mathematical ModelingInstitut for Informatik og Matematisk Modellering, 2014.

[20] Alain Girault and Hamoudi Kalla. A novel bicriteria scheduling heuristics providing a guaranteed global system failure rate. *IEEE Trans. Dependable and Secure Computing*, 6(4):241–254, Oct.–Dec. 2009.

[21] J. Golson. Tesla driver killed in crash with autopilot active, nhtsa investigating. Apr. 2016.

[22] Zonghua Gu, Gang Han, Haibo Zeng, and Qingling Zhao. Security-aware mapping and scheduling with hardware co-processors for flexray-based distributed embedded systems. *IEEE Trans. Parallel Distrib. Syst.*, PP:1–1, Oct. 2016.

[23] Zonghua Gu, Chao Wang, Ming Zhang, and Zhaohui Wu. Wcet-aware partial control-flow checking for resource-constrained real-time embedded systems. *IEEE Trans. Ind. Electron.*, 61(10):5652–5661, Oct. 2014.

[24] Chris Hobbs and Patrick Lee. Understanding iso 26262 asils. Jul. 2013.

[25] Menglan Hu, Jun Luo, Yang Wang, Martin Lukasiewycz, and Zeng Zeng. Holistic scheduling of real-time applications in time-triggered in-vehicle networks. *IEEE Trans. Ind. Informat.*, 10(3):1817–1828, May 2014.

[26] IEC. Iec 61508: Functional safety of electrical/electronic/programmable electronic safety-related systems. *International Electrotechnical Commission*, 2010.

[27] interfaces. http://copperhilltech.com/embedded-can-interfaces/.

[28] ISO. Iso 26262–road vehicles-functional safety. *International Organization for Standardization in ISO 26262*, 2011.

[29] ISO. Iso/pas 21448:2019-road vehicles–safety of the intended functionality. *International Organization for Standardization in ISO/PAS 21448.*, 2011.

[30] ISO. Road vehicles-functional safety, ISO 26262. *International Organization for Standardization in ISO 26262*, Dec. 2018.

[31] W. Jiang, P. Pop, and K. Jiang. Design optimization for security- and safety-critical distributed real-time applications. *Microprocessors Microsystems*, 52:401–415, 2016.

[32] Navneet Kaur and Sarbjeet Singh. A budget-constrained time and reliability optimization bat algorithm for scheduling workflow applications in clouds. *Procedia Computer Science*, 98:199–204, Jan. 2016.

[33] Amir Kazeminia. Reliability optimization of hardware components and system´s topology during early design phase. *Journal of American Association for Pediatric Ophthalmology and Strabismus*, 18(4):e9, 2014.

[34] V. Lesi, I. Jovanov, and M. Pajic. Network scheduling for secure cyber-physical systems. In *2017 IEEE Real-Time Systems Symposium (RTSS)*, 2017.

[35] Zhaojun Li, Mohammadsadegh Mobin, and Thomas Keyser. Multi-objective and multi-stage reliability growth planning in early product-development stage. *IEEE Trans. Rel.*, 65(2):769–781, Jun. 2016.

[36] Zheng Li, Li Wang, Shuhui Li, Shangping Ren, and Gang Quan. Reliability guaranteed energy-aware frame-based task set execution strategy for hard real-time systems. *J. Syst. and Softw.*, 86(12):3060–3070, Dec. 2013.

[37] Chung Wei Lin, Qi Zhu, Calvin Phung, and Alberto Sangiovanni-Vincentelli. Security-aware mapping for can-based real-time distributed automotive systems. In *International Conference on Computer-aided Design*, 2013.

[38] Man Lin, Yongwen Pan, Laurence T Yang, Minyi Guo, and Nenggan Zheng. Scheduling co-design for reliability and energy in cyber-physical systems. *IEEE Trans. Emerg. Topics Comput.*, 1(2):353–365, Dec. 2013.

[39] Rouhollah Mahfouzi, Amir Aminifar, Soheil Samii, Petru Petru Eles, and Zebo Peng. Security-aware routing and scheduling for control applications on ethernettsn networks. *ACM Transactions on Design Automation of Electronic Systems (TODAES)*, 2019.

[40] Z. Man, Y. Tao, A. Shaukat, S. Bran, O. Oscar, N. Roland, and I. Karmele. Specifying uncertainty in use case models. *Journal of Systems and Software*, 144, 2018.

[41] Saad Mubeen, Jukka Maki-Turja, and Martin Sjodin. Worst-case response-time analysis for mixed messages with offsets in controller area network. In *Proc. IEEE 17th Conf. Emerging Technologies and Factory Automation*, pages 1–10. IEEE, 2012.

[42] Nicolas Navet and Francoise Simonot-Lion. Automotive embedded systems handbook. *Automotive Embedded Systems Handbook*, 53(10):751–763, 2008.

[43] Yiannis Papadopoulos, Martin Walker, M-O Reiser, Matthias Weber, D Chen, Martin Törngren, David Servat, Andreas Abele, Friedhelm Stappert, H Lonn, et al. Automatic allocation of safety integrity levels. In *Proc. 1st Workshop Critical Automotive Applications: Robustness and Safety*, pages 7–10. ACM, 2010.

[44] David Parker, Martin Walker, Lujs Silva Azevedo, Yiannis Papadopoulos, and Rui Esteves Araújo. Automatic decomposition and allocation of safety integrity levels using a penalty-based genetic algorithm. In *Proc. Int. Conf. Industrial, Engineering and Other Applications of Applied Intelligent Systems*, pages 449–459. Springer, 2013.

[45] Pratyush Patel, Manohar Vanga, and Björn B Brandenburg. Timershield: Protecting high-priority tasks from low-priority timer interference (outstanding paper). In *2017 IEEE Real-Time and Embedded Technology and Applications Symposium (RTAS)*, pages 3–12. IEEE, 2017.

[46] Florian Pölzlbauer, Robert I Davis, and Iain Bate. Analysis and optimization of message acceptance filter configurations for controller area network (can). In *Proceedings of the 25th International Conference on Real-Time Networks and Systems*, pages 247–256. ACM, 2017.

[47] Juan M. Rivas, J. Javier Gutiérrez, J. Carlos Palencia, and Michael Gonzalez Harbour. Deadline assignment in edf schedulers for real-time distributed systems. *IEEE Trans. Ind. Informat.*, 26(10):2671–2684, Sep. 2015.

[48] Ernst Rolf. Formal performance analysis in automotive systems design - a rocky ride to new grounds. In *Proc. 23rd Int. Conf. Computer Aided Verification*. Springer, 2011.

[49] SAE. Sae j3016: Taxonomy and definitions for terms related to onroad motor vehicle automated driving systems j3016-201401. Jan. 2014.

[50] SAE. Sae j3051: Cybersecurity guidebook for cyber-physical vehicle systems. *Society of Automotive Engineers*, Jan. 2016.

[51] S. M. Shatz and J. P. Wang. Models and algorithms for reliability-oriented task-allocation in redundant distributed-computer systems. *IEEE Trans. Rel.*, 38(1):16–27, Apr. 1989.

[52] J. Shuttleworth. Sae standards news: J3016 automated-driving graphic update. Jan. 2019.

[53] Poonam Singh, Maitreyee Dutta, and Naveen Aggarwal. Bi-objective hwdo algorithm for optimizing makespan and reliability of workflow scheduling in cloud systems. In *2017 14th IEEE India Council International Conference (INDICON)*, pages 1–9. IEEE, 2017.

[54] Dominik Sojer, Christian Buckl, and Alois Knoll. Propagation, transformation and refinement of safety requirements. In *Proc. 3rd Workshop Non-functional System Properties in Domain Specific Modeling Languages*, 2010.

[55] H Tabani, L. Kosmidis, J. Abella, and F. J. Cazorla. Assessing the adherence of industrial autonomous driving software to iso-26262 guidelines for software. In *56th Design Automation Conference (DAC)*, 2019.

[56] Domiţian Tămaş-Selicean and Paul Pop. Design optimization of mixed-criticality real-time embedded systems. *ACM Trans. Embedded Comput. Syst.*, 14(3):50, May 2015.

[57] Zhuo Tang, Ling Qi, Zhenzhen Cheng, Kenli Li, Samee U. Khan, and Keqin Li. An energy-efficient task scheduling algorithm in dvfs-enabled cloud environment. *J. Grid Comput.*, 14(1):55–74, Mar. 2016.

[58] Haluk Topcuoglu, Salim Hariri, and Min-you Wu. Performance-effective and low-complexity task scheduling for heterogeneous computing. *IEEE Trans. Parallel Distrib. Syst.*, 13(3):260–274, Mar. 2002.

[59] J. D. Ullman. Np-complete scheduling problems. *J. Comput. Syst. Sci.*, 10(3):384–393, Jun. 1975.

[60] G. Xie, H. Peng, Z. Li, J. Song, Y. Xie, R. Li, and K. Li. Reliability enhancement toward functional safety goal assurance in energy-aware automotive cyber-physical systems. *IEEE Trans. Ind. Informat.*, 2018.

[61] Guoqi Xie, Yuekun Chen, Yan Liu, Yehua Wei, Renfa Li, and Keqin Li. Resource consumption cost minimization of reliable parallel applications on heterogeneous embedded systems. *IEEE Trans. Ind. Informat.*, 13(4):1629–1640, Aug. 2017.

[62] Guoqi Xie, Junqiang Jiang, Yan Liu, Renfa Li, and Keqin Li. Minimizing energy consumption of real-time parallel applications on heterogeneous systems. *IEEE Trans. Ind. Informat.*, PP:1–1, Mar. 2017.

[63] Guoqi Xie, Gang Zeng, Yuekun Chen, Yang Bai, Zhili Zhou, Renfa Li, and Keqin Li. Minimizing redundancy to satisfy reliability requirement for a parallel application on heterogeneous service-oriented systems. *IEEE Trans. Services Comput.*, PP:1–1, Feb. 2017.

[64] Guoqi Xie, Gang Zeng, Ryo Kurachi, Hiroaki Takada, Zhetao Li, Renfa Li, and Keqin Li. Wcrt analysis of can messages in gateway-integrated in-vehicle networks. *IEEE Transactions on Vehicular Technology*, 66(11):9623–9637, Dec. 2017.

[65] Guoqi Xie, Gang Zeng, Zhetao Li, Renfa Li, and Keqin Li. Adaptive dynamic scheduling on multi-functional mixed-criticality automotive cyber-physical systems. *IEEE Trans. Veh. Technol.*, 66(8):6676–6692, Aug. 2017.

[66] Guoqi Xie, Gang Zeng, Liangjiao Liu, Renfa Li, and Keqin Li. Mixed real-time scheduling of multiple dags-based applications on heterogeneous multi-core processors. *Microprocess. Microsy.*, 47:93–103, Nov. 2016.

[67] Guoqi Xie, Gang Zeng, Liangjiao Liu, Renfa Li, and Keqin Li. High performance real-time scheduling of multiple mixed-criticality functions in heterogeneous distributed embedded systems. *J. Syst. Architect.*, 70:3–14, Oct. 2016.

[68] Y. Xie, G. Zeng, K. Ryo, T. Hiroaki, and G. Xie. Security/timing-aware design space exploration of can fd for automotive cyber-physical systems. *IEEE Transactions on Industrial Informatics*, PP(2):1094–1104, 2019.

[69] Juan Yi, Qingfeng Zhuge, Jingtong Hu, Shouzhen Gu, Mingwen Qin, and H. M. Sha. Reliability-guaranteed task assignment and scheduling for heterogeneous multiprocessors considering timing constraint. *J. Signal Processing Syst.*, 81(3):1–17, Dec. 2015.

[70] R. Zalman and A. Mayer. A secure but still safe and low cost automotive communication technique. In *Proc. ACM 51st Annu. Des. Autom. Conf.*, pages 1–5, 2014.

[71] Haibo Zeng, Marco Di Natale, Arkadeb Ghosal, and Alberto Sangiovanni-Vincentelli. Schedule optimization of time-triggered systems communicating over the flexray static segment. *IEEE Trans. Ind. Informat.*, 7(1):1–17, 2011.

[72] Haibo Zeng, Marco Di Natale, Paolo Giusto, and Alberto Sangiovanni-Vincentelli. Stochastic analysis of can-based real-time automotive systems. *IEEE Trans. Ind. Informat.*, 5(4):388–401, Sep. 2009.

[73] Baoxian Zhao, Hakan Aydin, and Dakai Zhu. On maximizing reliability of real-time embedded applications under hard energy constraint. *IEEE Trans. Ind. Informat.*, 6(3):316–328, Aug. 2010.

[74] Baoxian Zhao, Hakan Aydin, and Dakai Zhu. Shared recovery for energy efficiency and reliability enhancements in real-time applications with precedence constraints. *ACM Transact. Des. Autom. Electron. Syst.*, 18(2):99–109, Mar. 2013.

[75] Laiping Zhao, Yizhi Ren, and Kouichi Sakurai. Reliable workflow scheduling with less resource redundancy. *Parallel Computing*, 39(10):567–585, 2013.

[76] Laiping Zhao, Yizhi Ren, Yang Xiang, and Kouichi Sakurai. Fault-tolerant scheduling with dynamic number of replicas in heterogeneous systems. In *High Performance Computing and Communications (HPCC), 2010 12th IEEE International Conference on*, pages 434–441. IEEE, 2010.

[77] Qingling Zhao, Zaid Al-Bayati, Zonghua Gu, and Haibo Zeng. Optimized implementation of multirate mixed-criticality synchronous reactive models. *ACM Transactions on Design Automation of Electronic Systems*, 22(2):23, Dec. 2016.

[78] Zhu, Qi, Lin, Chung-Wei, Sangiovanni-Vincentelli, Alberto, and Zheng Bowen. Security-aware design methodology and optimization for automotive systems. *ACM Transactions on Design Automation of Electronic Systems*, 21(1), 2016.

[79] Dakai Zhu and Hakan Aydin. Reliability-aware energy management for periodic real-time tasks. *IEEE Trans. Comput.*, 58(10):1382–1397, Oct. 2009.